RENEWABLE ENERGY: RESEARCH, DEVELOPMENT AND POLICIES

APPLIED SOFT COMPUTING TECHNIQUES FOR RENEWABLE ENERGY

RENEWABLE ENERGY: RESEARCH, DEVELOPMENT AND POLICIES

Additional books and e-books in this series can be found on Nova's website under the Series tab.

RENEWABLE ENERGY: RESEARCH, DEVELOPMENT AND POLICIES

APPLIED SOFT COMPUTING TECHNIQUES FOR RENEWABLE ENERGY

AMIT KUMAR THAKUR
RAJESH SINGH
AJAY KUMAR KAVITI
ANITA GEHLOT
AND
J. V. MURUGA LAL JEYAN
EDITORS

nova
science publishers
New York

NOTICE TO THE READER

The Publisher has taken reasonable care in the preparation of this book, but makes no expressed or implied warranty of any kind and assumes no responsibility for any errors or omissions. No liability is assumed for incidental or consequential damages in connection with or arising out of information contained in this book. The Publisher shall not be liable for any special, consequential, or exemplary damages resulting, in whole or in part, from the readers' use of, or reliance upon, this material. Any parts of this book based on government reports are so indicated and copyright is claimed for those parts to the extent applicable to compilations of such works.

Independent verification should be sought for any data, advice or recommendations contained in this book. In addition, no responsibility is assumed by the Publisher for any injury and/or damage to persons or property arising from any methods, products, instructions, ideas or otherwise contained in this publication.

This publication is designed to provide accurate and authoritative information with regard to the subject matter covered herein. It is sold with the clear understanding that the Publisher is not engaged in rendering legal or any other professional services. If legal or any other expert assistance is required, the services of a competent person should be sought. FROM A DECLARATION OF PARTICIPANTS JOINTLY ADOPTED BY A COMMITTEE OF THE AMERICAN BAR ASSOCIATION AND A COMMITTEE OF PUBLISHERS.

Additional color graphics may be available in the e-book version of this book.

Library of Congress Cataloging-in-Publication Data

Names: Kumar Thakur, Amit, editor.
Title: Applied soft computing techniques for renewable energy / Amit Kumar Thakur, editor, Lovely Professional University, Phagwara, Punjab, India, Rajesh Singh, editor, Lovely Professional University, Jalandhar, Punjab, India, Ajay Kumar Kaviti, editor, VNRJIET, Hyderabad, India, Anita Gehlot, editor, Lovely Professional University, Jalandhar, Punjab, India, J.V Muruga Lal Jeyan, editor, Lovely Professional University, Jalandhar, Punjab, India.
Description: Hauppauge, New York : Nova Science Publishers, 2020. | Series: Renewable energy: research, development and policies | Includes bibliographical references and index. |
Identifiers: LCCN 2020026264 (print) | LCCN 2020026265 (ebook) | ISBN 9781536181807 (hardcover) | ISBN 9781536182095 (adobe pdf)
Subjects: LCSH: Renewable energy sources--Data processing. | Artificial intelligence--Industrial applications.
Classification: LCC TJ808 .A67 2020 (print) | LCC TJ808 (ebook) | DDC 621.0420285/63--dc23
LC record available at https://lccn.loc.gov/2020026264
LC ebook record available at https://lccn.loc.gov/2020026265

Published by Nova Science Publishers, Inc. † New York

CONTENTS

PREFACE

Energy is of major importance to civilization, because it is driving force which binds human race. The estimation of energy in the form of renewable and sustainable is one of the important aspects to understand the how resources are harnessed and to predict what might happen under various possible future conditions.

This book aims in exploring the available modelling techniques to generate the best algorithms, which helps to determine the best solution in terms of comparing the performances of the solutions through different parameters for a specific case. This focuses on to carry out the comprehensive review and various other experimental works of various researchers who have carried out the work on these various soft computing techniques on various engines with various alternative fuels and other related technologies.

The book comprises of total nine chapters.

Chapter 1 describes a rigorous hybrid optimization approach for optimal design of reconfigurable microgrids (MGs). The proposed approach uses Mouth Flame Optimization (MFO) algorithm for optimal searching location of DGs and reconfiguration, and heuristic fuzzy has been integrated with MFO for effective handling of multi-objective.

Chapter 2 describes Renewable energy production from Biomass. It also explains the biomass based energy in carbon markets and the new technologies for the present day biomass energy.

Chapter 3 reveals the potential of hydrogen as an efficient energy carrier.

Chapter 4 provides safety related issues in handling and storage of gaseous hydrogen. It also gives an overview and discusses the key health and technical problems relevant to the hydrogen storage choices.

Chapter 5 describes widespread review of over 14 years of literature on the use of diesel-butanol fuel in diesel engines and its impact on combustion, productive use and emissions.

In Chapter 6 specific soft computing strategies for evaluating the performance and emissions of a spark ignition engine using alcohol-gasoline blend are discussed in details

Chapter 7 illustrates energy conservation techniques for food processing and manufacturing industries. As significant energy consumption in the food industry is due to electricity, heat, and cooling process so this excess energy intake can be reduced by implementing good energy-saving practices, which is described in this chapter.

Chapter 8 analyzes biodiesel output and pollution from various feedstocks. It also aims to complete an analysis of the strength, emissions and combustion of motors and their mixtures by biodiesel.

Chapter 9 provides a systematic way to understand the climate change, adoption and mitigation measures and climate finance and also emphasis on the actions taken at individual, organizational, societal, governmental level to preserve over Mother Nature.

ACKNOWLEDGMENTS

Editors are thankful to all the contributors and publisher for their support.

Dr. Amit Kumar Thakur, Lovely Professional University, India
Dr. Rajesh Singh, Lovely Professional University, India
Dr. Ajay Kumar Kaviti, VNR Vignana Jyothi Institute of Engineering & Technology, India
Dr. Anita Gehlot, Lovely Professional University, India
Dr. JV Muruga Lal Jeyan, Lovely Professional University, India

ABBREVIATIONS

ANR	Ammonia flow to Nitrous Oxide
aTDC	after Top Dead Centre
BTE	Brake Thermal Efficiency
BTE	Brake Thermal Efficiency
CA	Crank Angle
CC	Continuous Carburetion
CNG	Compressed Natural Gas
CO	Carbon monoxide
CO_2	Carbon dioxide
DCI	Direct Cylinder Injection
DEE	Di-Ethyl Ether
H2	Hydrogen
HC	Hydrocarbon
HCCI	Homogeneous Charge Compression Ignition
HCV	Higher Calorific Value
IC Engine	Internal Combustion Engine
IMEP	Indicated Mean Effective Pressure
LCV	Lower Calorific Value
LNT	Lean NOx Trap
LPG	Liquid Petroleum Gas
MFB	Mass Fraction Burned

MPI	Manifold Induction or Injection
PME	Palm Methyl Ester
SCR	Selective Catalytic Reduction
SVO	Straight Vegetable Oil
TDC	Top Dead Centre
TPI	Timed Port Injection
ULSD	Ultra-Low Sulphur Diesel

In: Applied Soft Computing Techniques … ISBN: 978-1-53618-180-7
Editors: Amit Kumar Thakur et al.

Chapter 1

INVESTIGATION OF WIND AND PHOTOVOLTAIC ENERGY SYSTEMS FOR DAILY LOAD DISPATCH ON RECONFIGURABLE MICROGRID THROUGH HYBRID FUZZY-MFO

***S. Thiruvenkadam*[1,2,*]*, Ravi Sankar*[3] *and In-Ho Ra*[2]

[1]P.A. College of Engineering and Technology,
Tamil Nadu, India
[2]School of Computer, Information and Communication Engineering,
Kunsan National University,
Gunsan, South Korea
[3]Department of Electrical Engineering,
University of South Florida,
Tampa, Florida, US

* Corresponding Author's Email: drstv.2011@gmail.com.

ABSTRACT

The microgrid (MG) is a power generation and distribution system composed of Distributed Generating (DG) units, Energy Storage Systems (ESSs), distributed reactive sources (DRS), and resilient loads, which can be operated in both connected and isolated modes. Notably, the handling of uncontrollable DGs, i.e., Renewable Energy Sources (RES) such as Wind Energy Systems (WES) and Photovoltaic Energy systems (PVES), becomes challenging for MGs planner to make decisions. To tackle technical and economic knowledge uncertainties, this paper proposes a rigorous hybrid optimization approach for the optimal design of reconfigurable MGs. The proposed method uses Mouth Flame Optimization (MFO) algorithm for optimal searching location of DGs and reconfiguration, and heuristic fuzzy has been integrated with MFO for effective handling of multi-objective such as minimization of voltage deviation, minimization of generation cost and loss reduction. This proposed hybrid algorithm investigates the optimal planning of MG to enrich the quality of power, significantly increase customer savings as well as the welfare of the Distribution System Operator (DSO). The optimal structures are identified by solving three different cases using the proposed algorithm, thereby giving DSO flexibility in decision making. Standard IEEE 33 bus Radial Distribution System (RDS) transformed into a microgrid is employed for validation. The validation is carried out under a 24-hour daily load pattern with 24-hour load dispatching characteristic behavior of WES and PVES. The results prove the superiority of the microgrid structure over the existing ones.

Keywords: renewable energy sources, radial distribution system, optimization, whale optimization algorithm, power loss

1. INTRODUCTION

1.1. General Perspective

Under the context of global energy infrastructures, the power system is undergoing significant changes in all areas related to planning, configuration, operation, and control. Due to the improved awareness of environmental issues and deregulation of the power industry, the demand for deploying distributed generation (DG) sources and energy storage systems (ESS) is increasing rapidly, with more and more DG units being adopted and installed near the houses. DGs and ESSs deployed near the loads bring huge benefits, such as enhanced system reliability, improved energy quality, and reduced system loss, power generation costs, and environmental pollution. However, the diversification of this paradigm transformation requires the coordination and cooperation of distribution systems and control systems, which also brings some difficulties in planning, operating, and controlling energy systems. Power distribution companies are facing a crucial challenge in designing an economical and reliable distribution system.

However, considering the economic factor of the distribution system, one of the effective solutions is reconfiguration, which aims to improve the reliability, reduce the power loss, and stabilize the voltage deviation.

1.2. Literature Review

Recently, many management strategies, control technologies, and optimization methods related to microgrid have emerged one after another, which provide reliable technical support for microgrids. In [1], the progress of intelligent distribution technology was briefly reviewed, and the emerging technology and future development trends were investigated to support the active management of the distribution network. Different frameworks of microgrids are proposed in [2, 3, 4, 5] to help energy management, power distribution, DG placement, and selection of RESs. The hierarchical structures of microgrid control methods were described in [6, 7, 8, 9, 10], to handle the system control, energy storage, power distribution, and market behaviors.

One goal of energy distribution in microgrids is to reduce the costs and losses. A hybridization of Grey Wolf Optimizer (GWO) and Particle Swarm Optimizer (PSO) to minimize power losses within the distribution system by identifying the optimal size and location of distributed generators (DGs) while considering distribution system restructuring was presented in [11]. Reconfiguration technology is a method that supports reducing the real power losses of RDS. In reference [12], the authors reviewed different optimization techniques employed for RDS reconfiguration, optimal location, and sizing of DG that enables minimization of the power losses. However, in [13], the day-ahead benefit of a microgrid is considered as the fitness function, and the Particle Swarm Optimization (PSO) algorithm was employed to increase the microgrid benefits by considering load response and uncertainties of WES. Also, an algorithm was proposed in [14] to execute Optimal Generation Scheduling (OGS) and Distribution Feeder Reconfiguration (DFR) on a short term forecast basis, for reducing the total operating costs of microgrid under various operational constraints of the utility system. In [15], an improved hybrid fuzzy-genetic algorithm was proposed to solve the network reconfiguration problem.

For obtaining an optimal solution of reconstruction and reduction of energy losses in microgrids, the parameters and characteristics of the specific objects were utilized to form the mathematic functions, for establishing a model, and then transform the energy optimization problem into a mathematical one. In [16], a Multi-objective Optimal Dispatch Model of Microgrid (MODMG) under uncertainty was proposed by using Interval Optimization (IO) method. The simulation results based on the improved IEEE112 bus MG proves the effectiveness of the Chaotic Group Search Optimizer with Multiple Producers (CGSOMP) algorithm. In [17], to achieve the goal of minimizing the total fuel cost and real power loss of MG, a hybrid optimization technique Chaotic Binary PSO (CBPSO) based on fuzzy theory was proposed to improve the speed of convergence and better handling of operational constraints along with the objective function. In [18], and efficient scenario-based stochastic programming approach was proposed to minimize the economic and emission costs, which decomposes the problem into deterministic problems. To improve the performance of the algorithm, a new adaptive probability correction strategy was proposed in this stochastic optimization method. The detailed comparison of the literature which dealt with solving the microgrid optimization problem using different optimization techniques is shown in Table 1.

Table 1. Comparison of literature

	MT	DE	FC	PV	WT	EV	ESS	VD	Proposed algorithm	Mathematical complexity
[19]		√		√			√	√	Grey Wolf Optimization (GWO) algorithm	Middle
[20]	√		√	√	√	√		√	Depth-First Search (DFS) algorithm	Simple
[21]				√	√			√	A sub-gradient ascent-based distributed algorithm	Middle
[22]	√		√	√	√		√		Particle Swarm Optimization (PSO) algorithm	Middle
[23]				√	√	√	√	√	Genetic algorithm (GA)	Middle
[24]					√			√	Time-varying acceleration coefficients particle swarm optimization (TVAC-PSO) algorithm	Complex
[25]			√	√	√		√	√	Modified firefly algorithm (MFA)	Middle
[26]				√				√	Fuzzy Optimization, and particle Swarm Optimization (PSO) algorithms	Complex
[27]				√	√			√	Particle Swarm Optimization (PSO)	Simple
[28]				√	√	√	√	√	Discrete genetic algorithm (DGA)	Simple
[29]	√			√	√	√	√	√	Non-dominated Sorting Genetic Algorithm (NSGAII)	Middle
[30]		√							Genetic Algorithm (GA)	Simple
[31]	√		√	√	√		√		matrix real-coded genetic algorithm (MRC-GA)	Middle
[32]	√		√	√	√		√		Adaptive Modified Firefly Algorithm (AMFA)	Middle
[33]				√				√	Multi-objective Antlion Optimizer (MAO) algorithm	Middle
[34]	√		√	√	√			√	Multi-objective Genetic Algorithm (MGA)	Middle

MT: Micro Turbine; FC: Fuel Cell; DE: Diesel Engine; WT: Wind Turbine; EV: Electric Vehicles; ESS: Energy Storage System; VD: Voltage Deviation.

1.3. Motivation and Aims

In most of the research shown in Table 1, the optimization problem has been solved by considering optimal DG placement as the objective function. Also, many of the research works analyzed controllable DGs such as MT, FC, and DE, for solving the optimization problem. However, these DGs may provide a cost-effective optimal solution, but still, the emission cost of the DGs is unavoidable. The worldwide rate of increase in CO_2 emission between the years 1751 and 2017 is shown in Figure 1 [35], which states that the present situation is alarming when considering the global warming effects.

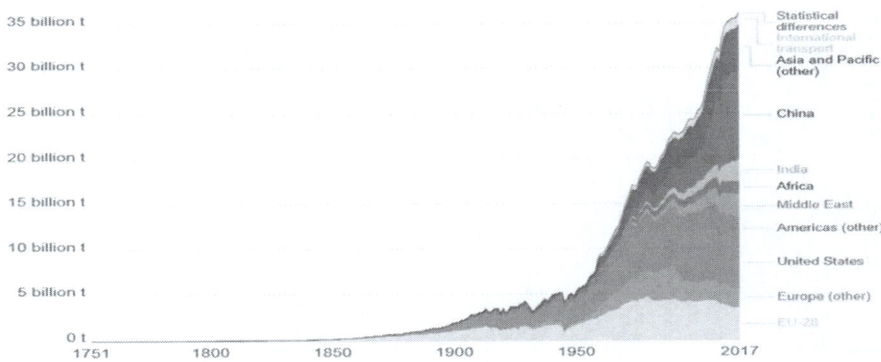

Figure 1. Worldwide CO_2 emission.

To control the impact of emission cost, the power generation through RES is most recommended to the utilities and utility customers. Most of the developing countries have realized the importance of limiting CO_2, NO_2, and SO_2emissions, which results in the replacement of RES for power production. Among all other RES, the power generations through WES and PVES are the most preferred based on various factors such as efficiency, lifetime, payback period, installation, and maintenance cost. Figure 2 reveals the world-wide year-by-year rate of increase in power generation through the RES [36]. However, it is clear from Figure 3 that the production rates of the WES and PVES still require significant improvement. This reflects the difficulty and uncertainties involved with the power generation through the two most popular ES. In this regard, much research is initiated with RES in a microgrid, which is listed in Table 1. However, the uncertainties involved with WES and PVES are unaddressed in many of the research. In this research, the impact of WES and PVES are analyzed using a hybrid Fuzzy and MFO algorithm for solving the distribution system optimization problem in the microgrid, considering their uncertainty behavior.

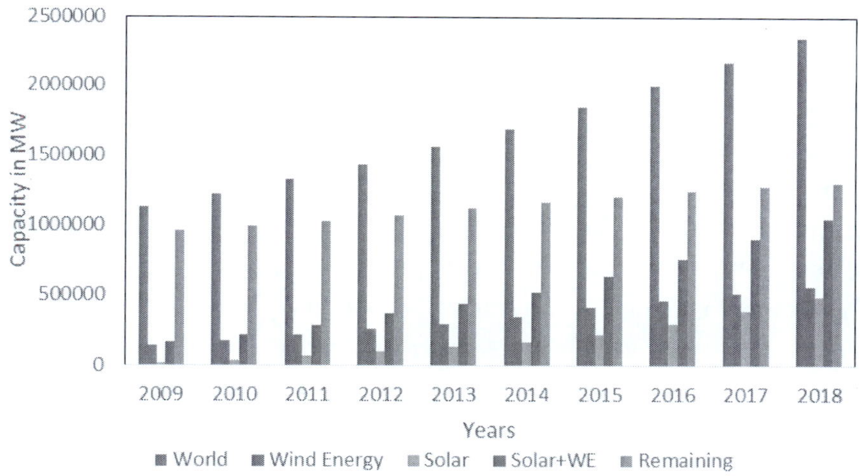

Figure 2. The worldwide growth rate of WES and PVES.

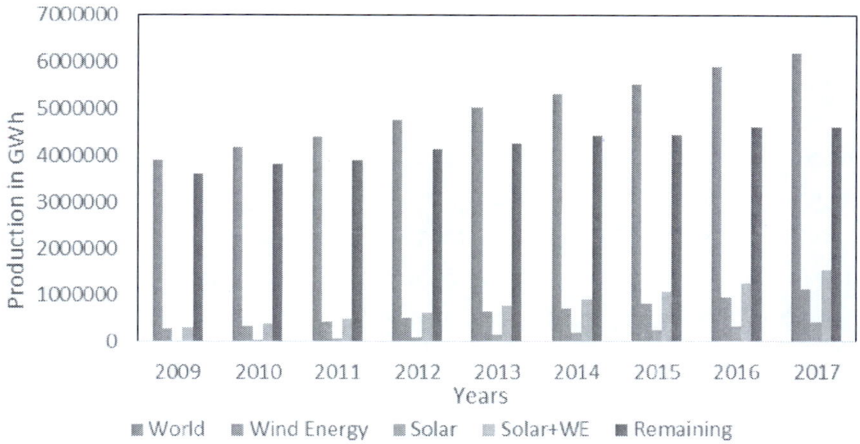

Figure 3. The worldwide production rate of WES and PVES.

2. MATHEMATICAL FORMULATION

The traditional distribution system has always been referred to as the passive end of the power transmission system, intending to provide consumers with electricity reliably and efficiently. To provide better protection to the utility customers, radial distribution systems are preferred at the customer side as the current flows in one direction, i.e., from the feeder to the read end. This well-established environment revolutionizes the greater adoption of DERs for performance enhancement of the system. The distribution system is undoubtedly not passive anymore. The ongoing research and development activities of the distribution system in numerous countries, i.e., the global distribution network, start with the partial realization of the smart grid. The active distribution system can be interconnected and subdivided into small units, regulated locally by the power controller, and power stream between local generators and loads. The adjacent units are also managed by the power controller. Information and communication technologies (ICTs) are commonly used in active grids to govern local communities, power generation, and absorption control. Also, it helps to control congestion by eliminating affected elements of the network and avoids the spread of overloads and failures.

In the present case, the DER connection conforms to the association, and the distributed generation is considered to be a combination of the negative load. Accordingly, only those generators that do not change the regular operation of the power grid can connect to the grid without grid investment. Taking into account the worst case in the power balance equation, i.e., the power imbalance concerning generation and demand, the DG effect is evaluated. In general, the lower the voltage level, the lower the connection cost, so the investment profitability of distributed power owners will be reduced. The connection and forgetting strategy pose severe restrictions in combining distributed power, especially in the weak RDS. This policy is unjust as it punishes the last owner of DG, who requires a connection. For these reasons, it should be recognized that the technical and economic obstacles caused by the passive operation of the distribution system still limit the use of DG and RE and is no longer acceptable to force the reduction of CO_2 emissions.

The design and operation practice of the distribution system is usually based on radial power flow, which brings unique challenges to the successful integration of DERs. To ensure that the distributed power supply will not reduce the power quality, security, or reliability of the distribution system, the problems that must be considered are well described in [37].

The distribution network in the power system is intimately involved in the challenge to integrate non-dispatchable energy (for example, PVES and WES), in which energy generation is entirely dependent on the environmental considerations such as weather, wind speed and time of the day. Among all other renewable energy sources, power generation through photovoltaic and wind turbines are more flexible. The photovoltaic solar panel systems may be mounted on commercial or residential roofs or near industrial or agricultural loads on the land. In contrast, small wind turbines are typical in rural areas. The following sub-sections describe the standard configuration details of the WES and PVES most commonly used in the distribution network.

2.1. Photovoltaic Energy System Modelling

The energy produced by the PV system depends directly on the irradiance and the angle of radiation in the solar PV cell. The meteorological conditions of the atmosphere determine the amount of irradiance and the generation variability of the PV.

Therefore, it can be assumed that there is a linear relationship between the global solar irradiance and the photovoltaic generation output. The following factors are considered for modeling:

- Parameters related to the superposition of cloud distribution models to the global solar radiation time series under clear atmospheric conditions;

- Parameters of environmental and electrical factors that may affect production; and
- The performance of the photovoltaic system as given in (1):

$$P_{PV_{lm}} = k_{cloud} \cdot k_{BOS} \cdot \frac{G_{lum}}{G_{stc}} \cdot P_{PV_r} \quad (1)$$

Wherek_{cloud}represents the reduction factor due to the clouds present, k_{BOS}represents the reduction factor that considers the physical infrastructure of the PV system other than the photovoltaic panels (wiring, switches, solar inverter) in real-time operation,G_{lum}represents the global solar irradiance for clear atmospheric conditions in a specific hour of a generic day in the year, G_{stc}represents the global solar irradiance (1000W/m^2), andP_{PV_r}Represents the rated power of the PV plant.

Numerous factors may influence the value ofk_{BOS}[38]:

- variation in solar irradiance,
- the temperature of the photovoltaic module,
- power consumption of the photovoltaic module,
- shading of the photovoltaic module,
- loss from cables, switchboards, and circuits,
- performance of the DC/AC inverter,
- performance of the transformer (if necessary).

A solarimeter or pyranometer usually is used for measuring solar radiation of the system. The relative geometry of cloud propagation in the photovoltaic power plant must be understood to calculate the power output of the photovoltaic system correctly. Certain assumptions are made to represent the various situations of photovoltaic plants, such as the total solar irradiance can be reduced by 90% on cloudy days, and the total solar irradiance can be reduced by 50% due to shading. The power generation cost of different varieties of PVES during recent years is shown in Figure 4 [36], which shows that the power production cost is reduced due to the new developments in material science.

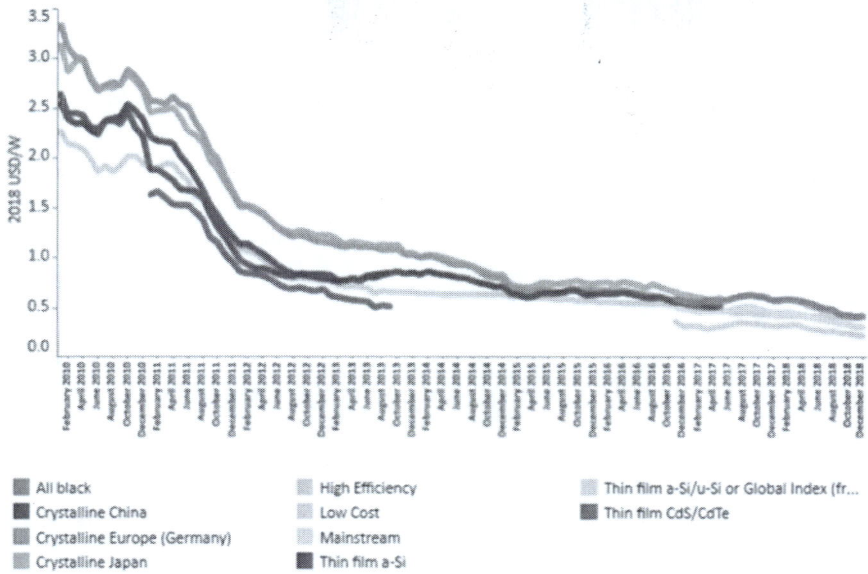

Figure 4. Power generation cost of different PVES.

2.2. Wind Energy System Modeling

In wind farms, power generation relies on the availability of primary power sources, which may vary in different time scales; the generation of wind energy is greatly affected by seasonal, regional, and time variations. At the same time, very short-term fluctuations within and between minutes must be taken into account, which is challenging to manage in weak distribution systems. The power generation of WES can be calculated by considering the time series of the main power supply and applying it to the generator model. Otherwise, the appropriate timing supply can be randomly formed according to the probability attribute of the main power supply. Starting from the time series, the nonlinear relationship given in (2) can be used to evaluate the output of a turbine.

$$P_{WT_h} = \begin{cases} 0 \; 0 \leq v_h \leq v_d \\ P_{WT_r} \frac{v_h - v_d}{v_r - v_d} \; v_d < v_h < v_r \\ P_{WT_r} \; v_r \leq v_h \leq v_\omega \\ 0 \; v_h > v_\omega \end{cases} \tag{2}$$

Here, v_d, v_ω and v_r are the cut-in, cut-off and rated wind speed of a wind plant, respectively, v_h is the actual wind speed, and P_{WT_r} is the installed capacity of the wind turbine. The power generation cost of different varieties of WES during recent years is shown in Figure 5 [36], which reveals that the power production cost for the Chinese turbines is much cheaper than other varieties of WES.

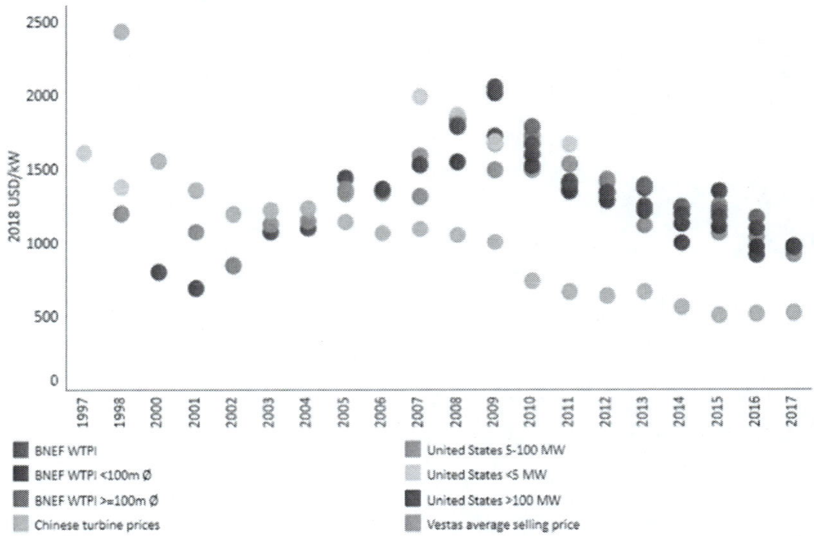

Figure 5. Power generation cost of different WES.

2.3. Objective Function

This optimization problem has been solved considering two objective functions, namely minimization of total generation cost and minimization of bus voltage deviation.

2.3.1. Minimization of Total Generation Cost

In this research, the optimization has been attempted with the implementation of restructuring and optimal DG placement. Therefore, the minimization of Total Generation Cost (TGC) is expressed as a function of Grid cost, WES cost, and PVES cost.

Minimize

$$TGC = Grid_{cost} + \sum_{i=1}^{nWES} WES_{cost,i} + \sum_{i=1}^{nPVES} PVES_{cost,i} \tag{3}$$

Where nWES and nPVES represent the total number of WES and PVES installed in the microgrid, respectively.

The cost function of the grid:

$$Grid_{cost} = P_{grid} * K_{grid} \tag{4}$$

Where K_{grid} represents the cost co-efficient constant, which is assumed as 0.044 $/kW-hr.

The cost function of WES [38]:

$$WES_{cost,i} = c_{WES,i}^{g} * \left(\frac{C_{WES,In,i}}{K_{WES,i}} + C_{WES,OM,i}\right) \tag{5}$$

Where $c_{WES,i}^{g}$ is an installed capacity of ith WES, $K_{WES,\ i}$ is a designed lifetime of ith WES, $C_{WES,In,i}$ *and* $C_{WES,OM,i}$ represents the installation, and operation and maintenance cost of the ith WES, respectively.

The cost function of PVES [38]:

$$PVES_{cost,i} = c_{PVES,i}^{g} * \left(\frac{C_{PVES,In,i}}{K_{PVES,i}} + C_{PVES,OM,i}\right) \tag{6}$$

Where $c_{PVES,i}^{g}$ is an installed capacity of *i*-th PVES, K_{PVES}, *i* isa designed lifetime of *i*-th PVES, $C_{PVES,In,i}$ *and* $C_{PVES,OM,i}$ represent the installation and operation and maintenance cost of the i-th PVES, respectively.

2.3.2. Minimization of Bus Voltage Deviation

To ensure the operational security and voltage quality to the customer ends, the minimization of bus voltage deviation is considered as an additional objective with the TGC reduction. The minimization of the bus voltage deviation with the set reference voltage for all the load buses of the system is expressed as,

$$VD = \sum_{i=1}^{nb} |V_i - V_{ref}| \tag{7}$$

Where V_{ref} is the reference voltage, V_i is the *i*-th bus voltage, and nb refers to the total number of load buses present in the system.

Constraints

This optimization problem has several constraints.

Equality Constraints

The constraints for power balance requirements:

$$\sum_{i=1}^{NG} PG_i - P_L = P_d \quad \sum_{i=1}^{NG} QG_i - Q_L = Q_d \tag{8}$$

Inequality Constraints

Generator Operation Constraints

The power generation limit of the generating units should be maintained within the range.

$$max(P_i^{min}, P_i^0 - DR_i) \leq P_i \leq min(P_i^{max}, P_i^0 + UR_i) \tag{9}$$

and

$$P_i \in \begin{cases} P_i^{min} \leq P_i \leq P_{i,1}^l \\ P_{i,j-1}^u \leq P_i \leq P_{i,j}^l \\ P_{i,n_i}^u \leq P_i \leq P_i^{max} \end{cases} ,j = 2,3,\ldots,n_i, i = l,\ldots,m. \tag{10}$$

Line Flow Constraints

The physical limits of electrical power equipment, such as capacity limits of generation units, power transformers, and reactive power sources, and apparent branch power flows.

$$P_{G_i}^{min} \leq P_{G_i} \leq P_{G_i}^{max} \quad 1 \leq i \leq N_G \tag{11}$$

$$\mathrm{Q_{G_i}^{min} \leq Q_{G_i} \leq Q_{G_i}^{max} \quad 1 \leq i \leq N_G} \tag{12}$$

$$\mathrm{Q_{C_i}^{min} \leq Q_{C_i} \leq Q_{C_i}^{max} \quad 1 \leq i \leq N_C} \tag{13}$$

$$\mathrm{|S_k| \leq S_k^{max} \quad 1 \leq k \leq N_E} \tag{14}$$

Bus voltage limits

$$0.95 \leq Vi \leq 1.05, i = 1,2,\ldots,nbus \tag{15}$$

DG Power factor limit

$$0.8 \leq PF \leq 1 \tag{16}$$

DG Limits

The real and reactive power handling capacities of the DG units are given below:

$$P_{gen,i,s} \leq P_{gen,\ i,max}, \quad \forall i \in N_{droop} \tag{17}$$

$$Q_{gen,i,s} \leq Q_{gen,i,max}, \quad \forall i \in N_{droop} \tag{18}$$

The real and reactive power handling capacities of the DG units are solved using a set of constraints given below:

$$0 \leq P_{gen,i,max} - P_{gen,\ i,s} \perp \frac{1}{m_{p,i}}(\omega_i^* - \omega_s) - P_{gen,\ i,s} \geq 0, \forall i \in N_{droop} \tag{19}$$

$$0 \leq Q_{gen\ ,i,max} - Q_{gen,\ i,s} \perp \frac{1}{n_{p,i}}\left(|V_k|^* - \left|V_{k,s}\right|\right)Q_{gen,i,s} \geq 0, \forall i \in N_{droop} \tag{20}$$

From the real and reactive power handling limits, the apparent power handling limits can be obtained as given below:

$$\sqrt{\left(\mathrm{P_{gen,i,s}}\right)^2 + \left(\mathrm{Q_{gen,i,s}}\right)^2} \leq \mathrm{S_{gen,i,max}}, \quad \forall \mathrm{i} \in \mathrm{N_{droop}} \tag{21}$$

Ramp Up/Down Rate Constraint

$$\begin{cases} \mathrm{P_m^{t,s}} - \mathrm{P_m^{t-1,s}} \leq \mathrm{r_m^{up}}\Delta \mathrm{t} \\ \mathrm{P_m^{t-1,s}} - \mathrm{P_m^{t,s}} \leq \mathrm{r_m^{down}}\Delta \mathrm{t} \end{cases} \tag{22}$$

Where $\mathrm{r_m^{up}}$ and $\mathrm{r_m^{down}}$ represent ramp up and down limits, respectively; $\Delta \mathrm{t}$ is the time period.

Storage Battery Constraints

$$\mathrm{E_{SB}^{t,s}} = \begin{cases} \mathrm{E_{SB}^{t-1,s}} - \mathrm{P_{SB}^{t,s}}\Delta \mathrm{t}\eta_{\mathrm{ch}} & \mathrm{P_{SB}^{t,s}} < 0 \\ E_{\mathrm{SB}}^{t-1,s} - P_{\mathrm{SB}}^{t,s}\Delta t/\eta_{\mathrm{dis}} P_{\mathrm{SB}}^{t,s} \geq 0 \end{cases} \tag{23}$$

$$P_{\mathrm{SB,ch}}^{max} \leq P_{\mathrm{SB}}^{t,s} \leq P_{\mathrm{SB,dis}}^{max} \tag{24}$$

$$E_{\mathrm{SB}}^{min} \leq E_{\mathrm{SB}}^{t,s} \leq E_{\mathrm{SB}}^{max} \tag{25}$$

Where $E_{\mathrm{SB}}^{t,s}$, E_{SB}^{min} and E_{SB}^{max} are the stored, minimum and maximum energy in SB, respectively, $P_{\mathrm{SB,ch}}^{max}$ and $P_{\mathrm{SB,dis}}^{max}$ are the charge and discharge rate of the SB; η_{ch} (η_{dis}) is the charge (discharge) efficiency.

3. Proposed Hybrid Algorithm

3.1. Moth-Flame Optimization Algorithm

Moth-flame Optimization (MFO) algorithm is a novel nature-inspired algorithm, which was first proposed by Seyedali Mirjalili in 2015 [40]. In the MFO algorithm, moths and flames are two key components, where moths are considered as the solutions calculated by randomly generated variables, and flames are regarded as the best solutions obtained so far. The moths can be described as a matrix:

$$M = \begin{bmatrix} M_{1,1} & M_{1,2} & \cdots & M_{1,d} \\ M_{2,1} & M_{2,2} & \cdots & M_{2,d} \\ \vdots & \vdots & \cdots & \vdots \\ M_{n,1} & M_{n,2} & \cdots & M_{n,d} \end{bmatrix} \tag{26}$$

Where n is the number of moths and d represents the number of variables. Accordingly, the fitness values of the moths can be described as follows:

$$OM = \begin{bmatrix} OM_1 \\ OM_2 \\ \vdots \\ OM_n \end{bmatrix} \tag{27}$$

Similarly, the flames have the same structure as the moths and can be described as follows:

$$F = \begin{bmatrix} F_{1,1} & F_{1,2} & \cdots & F_{1,d} \\ F_{2,1} & F_{2,2} & \cdots & F_{2,d} \\ \vdots & \vdots & \cdots & \vdots \\ F_{n,1} & F_{n,2} & \cdots & F_{n,d} \end{bmatrix} \tag{28}$$

Where n is the number of moths, and d is the number of variables. The corresponding fitness values are stored in a matrix:

$$OF = \begin{bmatrix} OF_1 \\ OF_2 \\ \vdots \\ OF_n \end{bmatrix} \tag{29}$$

The MFO algorithm is a 3-dimensional method for finding the global optimal solution of nonlinear problems, which can be defined as:

$$MFO = (I, P, T) \tag{30}$$

Where I is a function for producing random moths and their corresponding fitness values. The mathematical model for function I can be expressed as:

$$I: \emptyset \rightarrow \{M, OM\} \tag{31}$$

P is the main function, which moves moths around the search space, and records the final position through updating M, that is:

$$P: M \rightarrow M \tag{32}$$

T is the termination function, which returns true when the termination criterion is satisfied. Otherwise, the P function continues to search the space if the T returns false. T function is expressed as:

$$T: M \rightarrow \{true, false\} \tag{33}$$

After initialization, the P function iterates until the T function returns true. In this algorithm, a logarithmic spiral is defined to show the updating progress of moths:

$$S(M_i, F_j) = D_i e^{bt} cos(2\pi t) + F_j \quad (34)$$

$$D_i = |F_j - M_i| \quad (35)$$

Where D_i represents the distance between the i-th moth and the j-th flame, b is a constant for adjusting the shape of the logarithmic spiral, and t is a random value in $[r, 1]$ (r reduces linearly from -1 to -2 in the iterations).

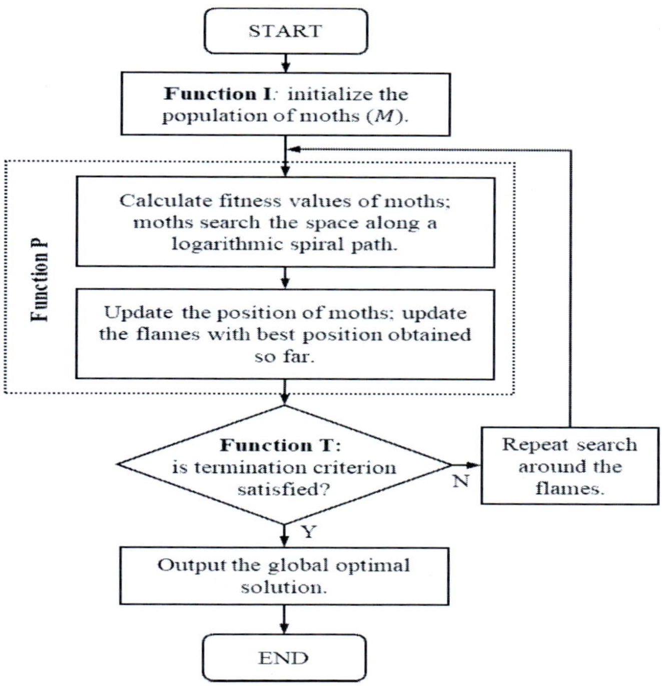

Figure 6. The flow chart of the MFO algorithm.

During iterations, the number of flames is decreasing gradually to enhance the detection of the optimal solution:

$$flameno = round(N - l\frac{N-1}{T}) \quad (36)$$

Where N is the number of moths, l represents the current number of iterations, and T is the maximum number of iterations.

The detailed flow chart explaining the iterative process of the MFO algorithm is shown in Figure 6 above.

The pseudocode of the proposed hybrid MFO algorithm is detailed below,

Initialization
N: number of moths
V: number of variables
F_{obj}: cost function
M: the matrix for moth and its variables
F: current optimal solutions (flames)
M_{pre}: the M in the previous iteration
OM_{pre}: the OM in the previous iteration
OM: fitness value of M
OF: fitness value of F
CI: current number of iterations
MI: maximum number of iterations
FN: number of flames
UB: upper bound of a variable
LB: lower bound of a variable
D: distance between M_i and F_j
Begin
For $CI = 1\ toMI$
$FN = round(N - CI * (N-1)/MI))$
For $i = 1\ to\ N$
$OM = sort(OM)$
For $j = 1\ to\ V$
$a = (M_{i,j} < LB_j), b = (M_{i,j} > UB_j)$
$M_{i,j} = M_{i,j} * (\sim(a+b)) + a * LB_j + b * UB_j$
End
$OM_i = F_{obj}(M_i)$
End
$M_{tmp} = M$
$OM_{tmp} = OM$
If $CI = 1$
$M = sort(M)$
$OM = sort(OM)$

$F = M$
$OF = OM$
Else
$M = [M_{pre}, M]$
$OM = [OM_{pre}, OM]$
$M = sort(M)$
$OM = sort(OM)$
$F = M, OF = OM$
End
$M_{pre} = M_{tmp}, OM_{pre} = OM_{tmp}$
For $i = 1$ to N
For $j = 1$ to V
If $i < FN$
$M_{i,j} = D \times e^{bt} \times cos(2\pi t) + F_{i,j}$
Else
$M_{i,j} = D \times e^{bt} \times cos(2\pi t) + F_{FN,j}$
End
End
End
Return optimal solution
End of algorithm

3.2. Heuristic Fuzzy

This optimization problem has two objective functions, namely the minimization of TGC and VD. Heuristic fuzzy has been incorporated to determine the compromised solution among these two objectives. The fuzzy membership functions are employed for each fitness function with maximum and minimum limits to give high flexibility for making decisions. The developed fuzzy functions can be modified according to the contribution of the fitness functions as per the problem needs. The detailed descriptions of the two objective functions are described below,

3.2.1. Membership Function for TGC

It is observed that the total generation cost of the solution should be minimized. Therefore, the ratio of total generation cost after optimization to distribution system generation cost before optimization for the given load level (TGC) is defined as follows:

$$TGC = \frac{TGC_{\text{after}}}{TGC_{\text{before}}} \tag{37}$$

In (37), the solution gets the maximum benefit for the lower values of TGC. The membership function corresponding to the minimization of TGC is expressed as,

$$\mu_{TGC} = \begin{cases} 1 \text{ TGC} \leq TGC_{\text{min}} \\ \frac{TGC_{\text{max}} - \text{TGC}}{TGC_{\text{max}} - TGC_{\text{min}}} TGC_{\text{min}} < TGC < TGC_{\text{max}} \\ 0 \; TGC \geq TGC_{\text{max}} \end{cases} \tag{38}$$

In this work, TGC_{min} and $TGC_{\text{max}}$are equal to 0.5 and 1, respectively. The valueTGC_{min}refers to the minimum acceptable amount of saving to 50%.

3.2.2. Membership Function for VD

Among the power flow constraints, bus voltage constraints are more sensitive when external equipment is connected to the system. Therefore, minimizing the bus voltage deviation has been considered as one of the objectives along with the TGC reduction. The minimum bus voltage of the system has been obtained from the load flow equation for each new population of FPOA. This membership function determines the deviation of the VD of the new population (VD_{aftert}) to the initial VD (VD_{before}). The membership function for the population can be expressed as follows,

$$\mu_{VD} = \begin{cases} 1.0 \; for \; VD_x \leq VD_{max} \\ \frac{VD_{max} - VD_x}{VD_{max} - VD_{min}} \; for \; VD_{min} < VD_x < VD_{max} \\ 0.0 \; for \; VD_x \geq VD_{max} \end{cases} \tag{39}$$

VD_{min} and VD_{max} are considered as 0.05 and 0.1, respectively. The VD_{min} reflects the maximum bus voltage of RDS is <1.05 (or the minimum bus voltage of RDS is >0.95), then $\mu_{VD} = 1$. The value of VD_x is calculated and expressed below,

$$VD_x = \frac{VD_{after}}{VD_{before}} \quad (40)$$

From the above equation, the fuzzified values for the TGC and VD can be obtained as μ_{TGC} and μ_{VD}, respectively. To obtain the compromise solution among these two fuzzified values, the fuzzy decision set is introduced which is expressed as,

$$\mu_D = \min\left(\mu_{TGC}, \mu_{VD}\right) \quad (41)$$

The equation (44) is considered as the fitness function for the MFO algorithm. As a result, each population of the MFO produces a compromised solution among the two objective functions.

4. Test System Data and Assumptions

In this work, the investigation is carried out on the standard IEEE 33 bus RDS. The single-line diagram of the test system is shown in Figure 7, which consists of 33 buses and 5 tie-lines. The selected base kV and MVA of the test system are 12.66kV, and 100MVA, respectively, and the total connected real and reactive power under base load conditions are 3.715MW and 2.3MVAR, respectively. The bus voltage minimum limit is assumed as 0.9 pu, and the maximum current carrying capacity of each line is expected to the maximum of 400A, excluding the line connecting the feeder. Under the reasonable operating condition, the NO's (Normally Open switches) are S_{33}, S_{34}, S_{35}, S_{36}, and S_{37}, and NC's (Normally Closed switches) are S_1 to S_{32}. The total transmission real power loss of the system under the baseload condition is 202.66 kW. The characteristic data of the test system with the line and load data is shown in Table 2. Also, the base price of energy purchased from the grid is assumed as 0.044$/kW-hr.

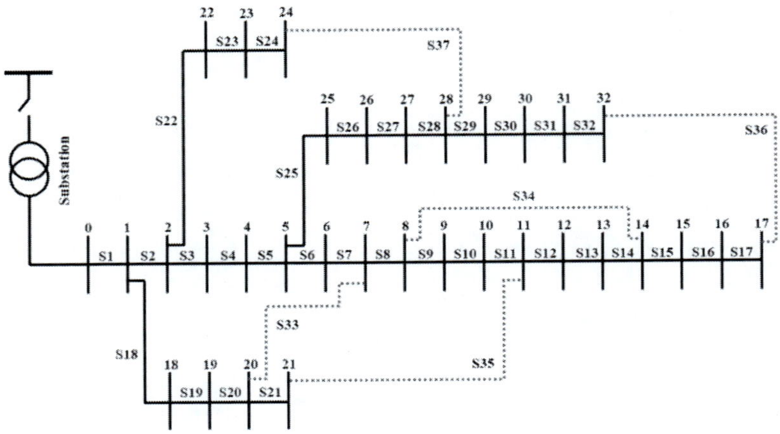

Figure 7. Single-line diagram of IEEE 33 Bus RDS.

Table 2. Network data of IEEE 33-bus test system

Bus To Bus	Line Resistance (ohm)	Line Reactance in(ohm)	Real Power (kW)	Reactive power (kVAR)	Bus To Bus	Line Resistance (ohm)	Line Reactance in(ohm)	Real Power (kW)	Reactive power (kVAR)
0-1	0.0922	0.0470	100	60	1-18	0.1640	0.1565	90	40
1-2	0.4930	0.2511	90	40	18-19	1.5042	1.3554	90	40
2-3	0.3660	0.1864	120	80	19-20	0.4095	0.4784	90	40
3-4	0.3811	0.1941	60	30	20-21	0.7089	0.9373	90	40
4-5	0.8190	0.7070	60	20	25-26	0.2842	0.1447	60	25
5-6	0.1872	0.6188	200	100	26-27	1.0590	0.9337	60	20
6-7	0.7114	0.2351	200	100	27-28	0.8042	0.7006	120	70
7-8	1.0300	0.7400	60	20	28-29	0.5075	0.2585	200	600
8-9	1.0440	0.7400	60	20	29-30	0.9744	0.9630	150	70
9-10	0.1966	0.0650	45	30	30-31	0.3105	0.3619	210	100
10-11	0.3744	0.1238	60	35	31-32	0.3410	0.5302	60	40
11-12	1.4680	1.1550	60	35	7-20	2	2		
12-13	0.5416	0.7129	120	80	8-14	2	2		
13-14	0.5910	0.5260	60	10	11-21	2	2		
14-15	0.7463	0.5450	60	20	17-32	0.5	0.5		
15-16	1.2890	1.7210	60	20	24-28	0.5	0.5		
16-17	0.7320	0.5740	90	40					

This work also addresses the optimization of a microgrid for the hour-to-hour dynamically varying active and reactive power demand. The assumed active and reactive power demands of the test system during each hour of a day is shown in Figure 8.

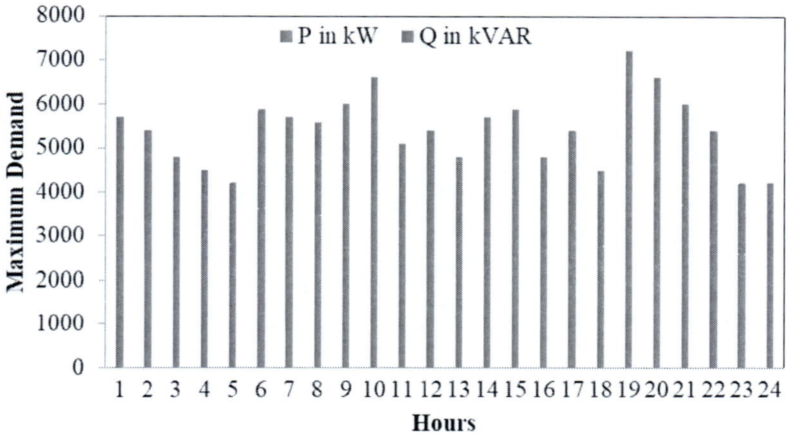

Figure 8. Power demands for the 24 hrs.

This study intends to analyze the impact of WES and PVES in the reconfigurable microgrid. The characteristic data of the two energy systems considered for the analysis are shown in Table 3. The hour-to-hour efficiency considered for the WES and PVES is shown in Figure 9. The details related to the energy systems are taken from [38].

Table 3. Characteristic data of PVES and WES

Parameters	Energy System	
	PVES	WES
Life Time	20	12.5
Installation Cost ($/kW)	7000	2500
Depreciation Time (in years)	20	10
Depreciation Cost ($/kW-year)	712.92	372.57
Installed Capacity (kW)	100	100

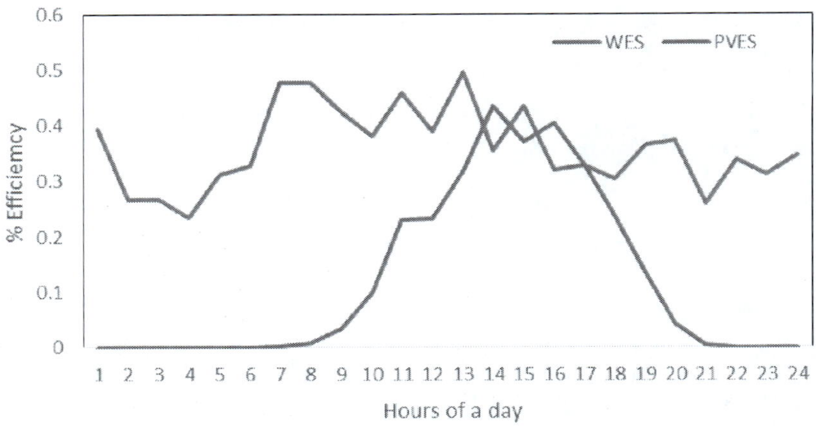

Figure 9. Percentage efficiency of ES in a day.

5. Simulation Results

Simulations have been carried out considering three different cases namely,

A. Performance investigation of microgrid with the presence of WES
B. Performance investigation of microgrid with the presence of PVES
C. Performance investigation of microgrid with the presence of both WES and PVES

In this work, the optimization involves both reconfiguration and optimal placement of RES. For the reconfiguration process, the common strategy has been adopted under all the mentioned test cases. The reconfiguration of an RDS is achieved through preparing the active zones. The active zone of a system is identified according to the number of normally-open switches present in the system as the tie-switches are most essential for restructuring. The test system considered for the simulation has five tie-switches hence resulting in five active zones. Figure 10 provides the zone wise splitting of switches.

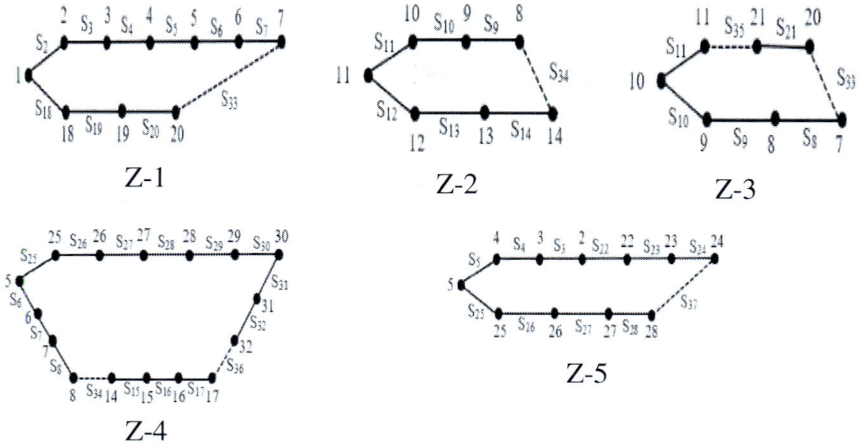

Figure 10. Active zones of IEEE 33 bus RDS.

The number of switches present in each zone is suggestively reduced when the individual zones are considered as solution sets for the reconfiguration. Nonetheless, still, the iterative system can avoid settling to the unfeasible solution sets. To neglect the unfeasible solution sets in the iterative process and improve the efficiency of the optimization, the switches present in the system are classified based on their location. The switches are labeled under four classifications such as Open, Closed, Permanently Closed, and Temporarily Closed [39]. The final set of switches in each zone is given in Table 4.

Table 4. Set of switches in each zone

Zone (Z)	Normally Closed Switches (NCs)	Normally Open Switches (NOs)
1	$\{S_4, S_5, S_6, S_7, S_{19}, S_{20}\}$	$\{S_{33}\}$
2	$\{S_{12}, S_{13}, S_{14}\}$	$\{S_{34}\}$
3	$\{S_8, S_9, S_{10}, S_{11}, S_{21}\}$	$\{S_{35}\}$
4	$\{S_{15}, S_{16}, S_{17}, S_{29}, S_{30}, S_{31}, S_{32}\}$	$\{S_{36}\}$
5	$\{S_{23}, S_{24}, S_{25}, S_{26}, S_{27}, S_{28}\}$	$\{S_{37}\}$

5.1. Performance Investigation of Microgrid with the Presence of WES

This test case has been carried out for investigating the performance of microgrid with the presence of WES alone in the RDS. The analysis is conducted with assuming only five WES is permitted to the microgrid. Therefore, the proposed hybrid algorithm keeps five variables for reconfiguration and another five variables for finding the optimal location of WES in the microgrid. The proposed algorithm is executed for 50 iterations and 50 populations.

After successful execution, buses 31 and 32 are identified as the optimal locations for WES placement. The identified NOs of the optimal configuration are S_7, S_9, S_{14}, S_{32}, and S_{37}. Figure 11 shows the optimal structure of the microgrid along with the WES location.

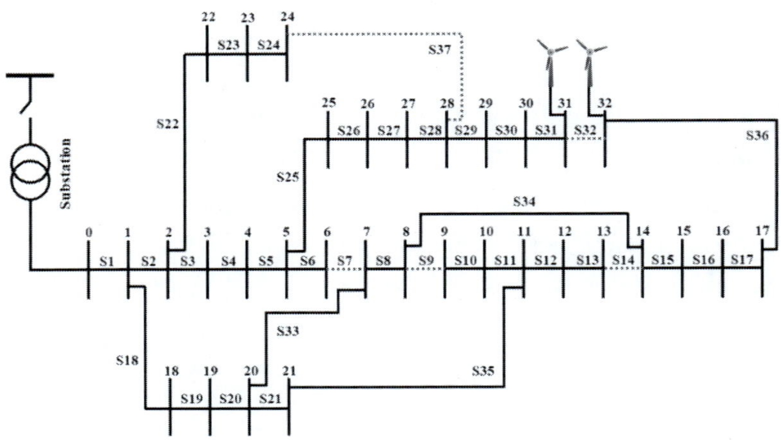

Figure 11. The optimal structure with WES.

The final configuration bus voltages with the presence of WES are compared with the initial configuration bus voltage shown in Figure 12. It details that the inclusion of the WES to the identified optimal location, which improves the system bus voltage compared to the initial configuration. Moreover, the bus voltages are maintained above 0.95 pu (i.e., it records 0.9535pu as the minimum bus voltage at the 32nd bus), which facilitates the system for further loading at the buses. Besides, the loading at the branches is shown in Figure 13, which ensures that the branch loading is reduced compared to the initial configuration. The transmission loss of the system is brought to 94.15 kW from its initial configuration loss of 202.66kW. The total generation cost of the system is 194.68$/hr with the presence of WES. From the obtained results, it is identified that the generation through the WES is effective and facilitates the system for further loading. Most importantly, the inclusion of the WES reduces the total emission cost

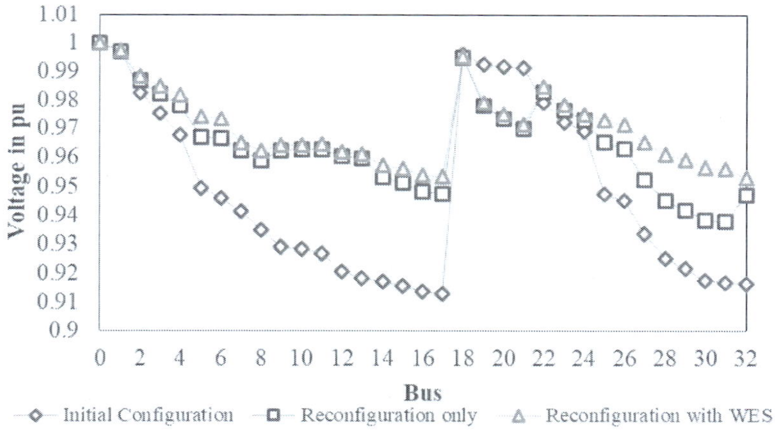

Figure 12. Comparison of bus voltages with WES under different configurations.

Furthermore, the algorithm convergence characteristics are compared with the other algorithms where the algorithms are executed for the same iteration and population values. Figure 14 illustrates that the proposed algorithm brings the optimum result at 35th iteration, and the speed of convergence to the optimum result is better than the other algorithms. Moreover, it reaches the near-optimal solution within the 20th iteration. The algorithms DGA [28] and PSO [26] fails to bring the optimal solution for the given number of iterations and populations. However, the optimal solutions are received by the DGA and PSO after increasing the iterations and populations.

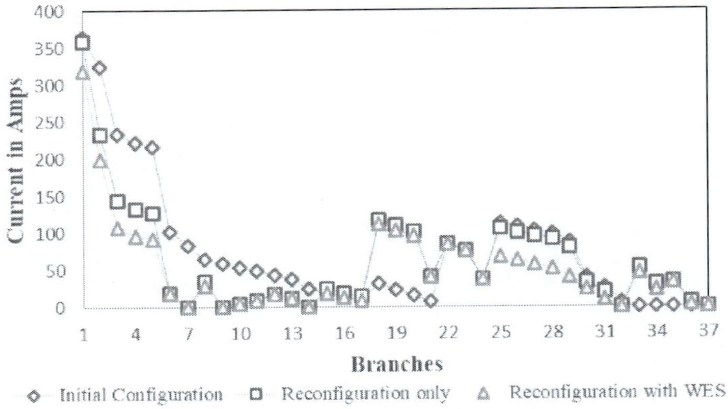

Figure 13. Comparison of branch currents with WES under different configurations.

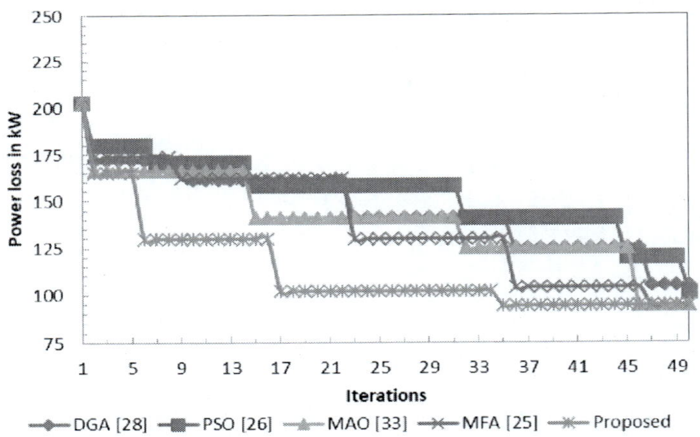

Figure 14. Convergence characteristic of the different algorithms.

The strength of the proposed hybrid algorithm is validated by keeping the test system working in variable demand. The typical load pattern on a day shown in Figure 8 and the varying hourly efficiency of the typical WES on a day shown in Figure 9 is used for the study. The optimal solution is obtained after successfully implementing the algorithm under varying hourly generation and load patterns, which is shown in Table 5. In most of the solutions summarized in Table 5, the identified optimal locations for WES are 16, 17, 31, and 32, and the NO's for the optimal structure as S_7, S_9, S_{14}, S_{32}, and S_{37}.

Table 5. Summary of results with WES under daily load pattern

Hrs.	% Efficiency of WES	Total Load	Configuration	Optimal WES location	P_{WES} in MW	P_{Grid} in MW	TGCin $/hr	Loss in kW	Min. Voltagein pu	Bus
1	0.364	3.5293	7,9,14,32,37	31,31,31,32,32	0.1820	3.4352	187.5549	87.9045	0.9526	32
2	0.267	3.3435	7,9,14,28,32	31,17,32,32,17	0.1335	3.2896	181.1488	79.5608	0.9560	32
3	0.267	2.9720	9,14,28,33,36	32,31,32,17,17	0.1335	2.8992	163.9749	60.7445	0.9647	17
4	0.234	2.7863	9,14,28,33,36	16,32,16,16,17	0.1170	2.7220	156.1776	52.7840	0.9670	17
5	0.312	2.6005	9,14,17,28,33	32,16,32,32,32	0.1560	2.4876	145.8632	43.1146	0.9718	17
6	0.329	3.6407	7,9,14,32,37	31,32,32,31,31	0.1645	3.5695	193.4682	93.3470	0.9522	32
7	0.476	3.5293	7,9,14,32,37	31,31,17,32,31	0.2380	3.3754	184.9260	84.1568	0.9541	32
8	0.477	3.4550	7,9,14,28,32	31,31,32,32,17	0.2385	3.2969	181.4738	80.4974	0.9575	32
9	0.424	3.7150	7,9,14,32,37	32,32,31,31,31	0.2120	3.5972	194.6830	94.1552	0.9535	32
10	0.381	4.0865	7,9,14,17,28	17,32,31,17,32	0.1905	4.0194	213.2629	123.4266	0.9482	16
11	0.459	3.1578	9,14,28,33,36	31,17,16,32,32	0.2295	2.9908	168.0031	62.5456	0.9642	17
12	0.390	3.3435	7,9,14,28,36	17,17,32,32,32	0.1950	3.2243	178.2790	75.8388	0.9602	17
13	0.494	2.9720	9,14,28,33,36	17,17,31,32,32	0.2470	2.7800	158.7290	55.0196	0.9676	17
14	0.355	3.5293	7,9,14,32,37	31,31,32,32,31	0.1775	3.4400	187.7665	88.2137	0.9525	32
15	0.433	3.6407	7,9,14,32,37	31,17,31,31,32	0.2165	3.5140	191.0230	89.7753	0.9535	32
16	0.321	2.9720	9,14,28,33,36	31,32,17,17,32	0.1605	2.8708	162.7243	59.3213	0.9654	17
17	0.329	3.3435	7,9,14,28,32	32,31,32,17,17	0.1645	3.2566	179.7004	77.6424	0.9577	32
18	0.303	2.7863	9,14,28,33,36	32,17,32,16,17	0.1515	2.6859	154.5867	51.1264	0.9672	17
19	0.364	4.4580	7,9,14,28,36	32,31,17,17,32	0.1820	4.4151	230.6737	139.1266	0.9463	17
20	0.373	4.0865	7,9,14,17,28	17,17,32,32,32	0.1865	4.0237	213.4523	123.7320	0.9482	16
21	0.260	3.7150	7,9,14,32,37	32,31,31,32,31	0.1300	3.6850	198.5474	99.9832	0.9511	32
22	0.338	3.3435	7,9,14,28,32	32,32,17,17,31	0.1690	3.2519	179.4905	77.3720	0.9580	32
23	0.312	2.6005	9,14,17,28,33	32,17,32,16,17	0.1560	2.4876	145.8624	43.0976	0.9720	16
24	0.346	2.6005	9,14,17,28,33	32,32,31,16,17	0.1730	2.4699	145.0840	42.4053	0.9723	16

5.2. Performance Investigation of Microgrid with the Presence of PVES

This test case is executed by considering only that the PVES are present in the microgrid. Like the previous test case, it is assumed that only five PVES are available for power generation. The optimization process is carried out by considering the same boundary conditions which are used in the previous case for the proposed hybrid algorithm. The specific details of the PVES are referred from Table 3. After the successful implementation of the proposed algorithm, the optimal solution has arrived at the 33rd iteration. Results are obtained for the microgrid with the presence of PVES. The final configuration results are shown in Table 6. The reduction in power loss from the initial configuration loss, the optimal configuration, the optimal locations of PVES, and the minimum voltage of the configuration are detailed in Table 6.

Table 6. Optimal solution with PVES under base load condition

Optimal Configuration	PVES optimal location	Total Cost in $/hr	Power Loss in kW	Min Voltage in pu.	Bus Number
7,9,14,32,37	31,31,31,31,32	235.06	108.67	0.9476	32

The obtained optimal configuration bus voltages and branch currents are compared with the other configurations of the system, which are shown in Figure 15 and Figure 16, respectively. According to the figures, the bus voltages are improved, and the branch's current loading is reduced compared to the remaining two configurations.

The optimization process has been executed by considering the daily load pattern and the daily efficiency characteristics of PVES shown in Figure 8 and Figure 9, respectively. Table 7 summarizes the obtained results under variable load conditions. From Table 7, it is identified that in most of the obtained solutions, the optimal locations for PVES are 16, 17, 18, 31, and 32, and the NO's for the optimal structure as S_7, S_9, S_{14}, S_{32}, and S_{37}. Under varying conditions, it is recommended to connect the PVES at the identified locations for optimal solutions.

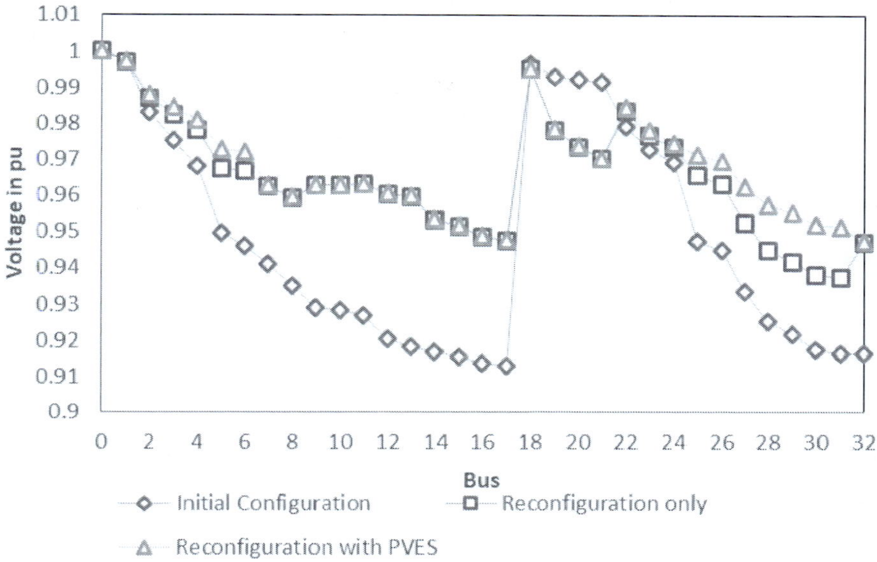

Figure 15. Comparison of bus voltages with PVES under different configurations.

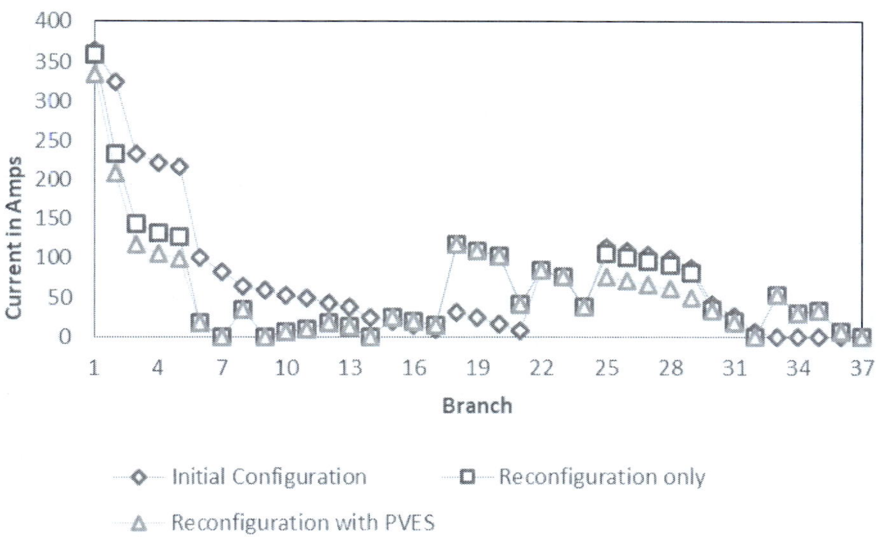

Figure 16. Comparison of branch currents with PVES under different configurations.

Table 7. Summary of results with PVES under daily load pattern

Hrs.	% Efficiency of PVES	Total Load in MW	Configuration	Optimal WES location	P_{PVES} in MW	P_{Grid} in MW	TGC in $/hr	Loss in kW	Min. Voltage in pu	Bus
1	0.000	3.5293	7,9,14,28,36	29,27,32,32,17	0.0000	3.6304	227.3270	101.1755	0.9523	32
2	0.000	3.3435	7,9,14,28,36	32,32,32,16,11	0.0000	3.4319	218.5910	88.3783	0.9549	17
3	0.000	2.9720	9,14,28,33,36	18,17,1,26,16	0.0000	3.0404	201.3638	68.3517	0.9613	17
4	0.000	2.7863	9,14,28,33,36	29,1,30,12,1	0.0000	2.8453	192.7812	59.0420	0.9618	17
5	0.000	2.6005	9,14,17,28,33	24,13,10,29,8	0.0000	2.6508	184.2233	50.2951	0.9672	17
6	0.000	3.6407	7,9,14,32,37	32,32,10,32,32	0.0000	3.7465	232.4324	105.7571	0.9474	32
7	0.002	3.5293	7,9,14,28,36	32,32,32,32,32	0.0010	3.6293	227.2796	101.0982	0.9524	32
8	0.008	3.4550	7,9,14,28,36	32,32,32,32,32	0.0040	3.5471	223.6625	96.1910	0.9541	32
9	0.035	3.7150	7,9,14,32,37	31,31,31,31,32	0.0175	3.8062	235.0599	108.6726	0.9476	32
10	0.100	4.0865	7,9,14,28,36	17,17,16,17,32	0.0500	4.1714	251.1311	134.9264	0.9421	17
11	0.230	3.1578	9,14,28,33,36	17,32,17,16,32	0.1150	3.1114	204.4917	68.6907	0.9626	17
12	0.233	3.3435	7,9,14,28,32	32,32,17,17,32	0.1165	3.3076	213.1236	80.6191	0.9567	32
13	0.318	2.9720	9,14,28,33,36	31,17,32,17,32	0.1590	2.8724	193.9739	59.3993	0.9654	17
14	0.433	3.5293	7,9,14,32,37	31,31,32,17,31	0.2165	3.3983	217.1146	85.5739	0.9535	32
15	0.370	3.6407	7,9,14,32,37	31,32,31,32,31	0.1850	3.5476	223.6836	91.9197	0.9528	32
16	0.403	2.9720	9,14,28,33,36	32,17,32,31,17	0.2015	2.8277	192.0086	57.2335	0.9665	17
17	0.330	3.3435	7,9,14,28,32	32,32,31,17,17	0.1650	3.2561	210.8573	77.6123	0.9578	32
18	0.238	2.7863	9,14,28,33,36	16,32,17,17,16	0.1190	2.7199	187.2650	52.6733	0.9672	17
19	0.133	4.4580	7,9,14,28,36	32,32,17,17,32	0.0665	4.5404	267.3662	148.9057	0.9431	17
20	0.043	4.0865	7,9,14,28,36	16,32,17,17,17	0.0215	4.2024	252.4961	137.4499	0.9405	17
21	0.003	3.7150	7,9,14,32,37	31,31,31,31,31	0.0015	3.8235	235.8211	109.9725	0.9474	32
22	0.000	3.3435	7,9,14,28,36	1,11,32,24,14	0.0000	3.4319	218.5910	88.3783	0.9549	17
23	0.000	2.6005	9,14,17,28,33	1,11,32,32,12	0.0000	2.6508	184.2233	50.2951	0.9672	17
24	0.000	2.6005	9,14,17,28,33	29,23,1,1,13	0.0000	2.6508	184.2233	50.2951	0.9672	17

5.3. Performance Investigation of Microgrid with the Presence of WES and PVES

As per this test case, the different combinations of WES and PVES are investigated by restricting a total of five numbers of ES for power generation. The optimization has been carried out with the proposed algorithm by keeping a population size of 50 and maximum iteration as 100. The investigation is done under different load demands of a day by considering the equivalent efficiency of WES and PVES in a particular hour.

Table 8. Summary of real power loss at the final configuration

Hrs.	(WES:4, PVES:1)	(WES:3, PVES:2)	(WES:2, PVES:3)	(WES:1, PVES:4)
1	90.4450	93.0794	95.7262	98.3647
2	81.2321	82.9850	84.7318	86.5383
3	62.1899	63.6802	65.1886	66.7549
4	53.9276	55.1014	56.3435	57.6568
5	44.4257	45.8211	47.2326	48.7218
6	95.7087	98.1131	100.5823	103.1584
7	87.3356	90.6370	94.0366	97.4709
8	83.3870	86.4000	89.5586	92.7972
9	96.8975	99.6706	102.5826	105.5833
10	125.6109	127.8717	130.1807	132.5208
11	63.7022	64.8959	66.1525	67.3974
12	76.7658	77.6960	78.6631	79.6365
13	55.8670	56.7357	57.6054	58.4973
14	87.6753	87.1453	86.6156	86.1035
15	90.2002	90.6263	91.0551	91.4870
16	58.8968	58.4769	58.0586	57.6455
17	77.6488	77.6304	77.6244	77.6183
18	51.4278	51.7440	52.0539	52.3654
19	141.0064	142.9366	144.8914	146.8809
20	126.3178	129.0053	131.7361	134.5194
21	101.9069	103.8470	105.8517	107.8988
22	79.4729	81.5796	83.7909	86.0674
23	44.4257	45.8197	47.2326	48.7218
24	43.8336	45.3610	46.9191	48.5554

Table 9. Summary of TGC at the final configuration

Hrs.	(WES:4, PVES:1)	(WES:3, PVES:2)	(WES:2, PVES:3)	(WES:1, PVES:4)
1	195.5043	203.4579	211.4120	219.3657
2	188.6332	196.1211	203.6088	211.0992
3	171.4493	178.9257	186.4029	193.8827
4	163.4936	170.8109	178.1312	185.4546
5	153.5297	161.1999	168.8709	176.5452
6	201.2557	209.0452	216.8375	224.6345
7	193.3875	201.8544	210.3257	218.7984
8	189.9006	198.3328	206.7714	215.2136
9	202.7513	210.8209	218.8967	226.9764
10	220.8314	228.4034	235.9774	243.5528
11	175.2977	182.5938	189.8928	197.1912
12	185.2467	192.2144	199.1838	206.1535
13	165.7767	172.8254	179.8741	186.9238
14	193.6357	199.5052	205.3747	211.2450
15	197.5550	204.0870	210.6191	217.1513
16	168.5808	174.4376	180.2944	186.1515
17	185.9323	192.1631	198.3945	204.6259
18	161.1220	167.6579	174.1936	180.7294
19	238.0089	245.3462	252.6847	260.0246
20	221.2541	229.0604	236.8686	244.6791
21	205.9989	213.4511	220.9061	228.3630
22	187.3062	195.1221	202.9426	210.7661
23	153.5297	161.1999	168.8709	176.5452
24	152.9052	160.7309	168.5579	176.3883

The proposed hybrid algorithm brings the optimal solution for all the combinations with different characteristic conditions. The optimal results are collected under varying load conditions for various combinations of WES and PVES. The real power loss, total generation cost, and minimum bus voltage of microgrid are shown in Tables 8, 9, and 10, respectively.

The following observations are obtained from the investigations,

i. The optimal configurations are almost the same under all the conditions of a day for many of the combinations of ES.

ii. The optimal locations for the ES are also almost the same irrespective of the type of ES, i.e., WES and PVES.

Table 10. Summary of minimum bus voltage at the final configuration

Hrs.	(WES:4, PVES:1)	(WES:3, PVES:2)	(WES:2, PVES:3)	(WES:1, PVES:4)
1	0.9526	0.9500	0.9526	0.9500
2	0.9560	0.9567	0.9567	0.9567
3	0.9645	0.9631	0.9629	0.9625
4	0.9660	0.9659	0.9645	0.9632
5	0.9708	0.9699	0.9697	0.9685
6	0.9522	0.9498	0.9498	0.9498
7	0.9541	0.9509	0.9508	0.9508
8	0.9577	0.9542	0.9541	0.9546
9	0.9507	0.9510	0.9504	0.9484
10	0.9482	0.9482	0.9446	0.9440
11	0.9644	0.9643	0.9629	0.9629
12	0.9584	0.9573	0.9573	0.9562
13	0.9674	0.9665	0.9665	0.9661
14	0.9531	0.9530	0.9535	0.9530
15	0.9535	0.9531	0.9531	0.9528
16	0.9659	0.9659	0.9660	0.9664
17	0.9576	0.9578	0.9578	0.9578
18	0.9672	0.9670	0.9680	0.9676
19	0.9448	0.9447	0.9438	0.9431
20	0.9482	0.9481	0.9445	0.9428
21	0.9493	0.9493	0.9492	0.9475
22	0.9557	0.9557	0.9572	0.9571
23	0.9708	0.9698	0.9697	0.9685
24	0.9713	0.9701	0.9700	0.9686

iii. The inclusion of RES dramatically reduces the emission cost by load sharing with the grid.

iv. The total generation cost of the system and real power loss decreases with the increase in the number of WES as the generation cost of WES is lesser compared to PVES.

v. The power generation by PVES is zero for nearly 12 hours in a day, and only for 8 hours of a day, PVES has more than 25% efficiency. However, during the summer seasons, the efficiency of PVES would be much better than WES.

vi. The per-day efficiency of WES is better compared with the PVES.

vii. Considering the lifetime, PVES are better compared with the WES.

viii. The PVES can be installed at any location as it does not have any specific environmental constraints, but WES installation requires the suitable wind velocity at the location.

CONCLUSION

The microgrid is not only the focus of current research but also brings more significant opportunities for the utilization of renewable energy resources. This chapter mainly discusses the microgrid reconfiguration after the addition of renewable energy resources in the current microgrid. To reduce the cost of microgrid like the reduction of the emission cost and provision of stable voltage, the MFO algorithm is used to optimize and reconfigure the location of DGs, and the fuzzy method is used to deal with multi-objective issues. This research investigated three different cases, i.e., the performance of microgrid only with WES, only with PVES, and combined WES and PVES. Furthermore, the load dispatch under hour-to-hour changing daily pattern was studied with the 24-hour characteristics of ES. The simulation results show that the proposed method ensures the optimal solution, and the optimized structure is better than the existing structure. In future works, more objectives will be considered, and a variety of novel optimization methods can be applied for obtaining a better reconfiguration solution in the microgrid. Also, the effect of other controllable DGs, along with renewable energy sources, can be studied. The answer to the microgrid under abnormal conditions can also be analyzed with DGs.

REFERENCES

[1] Zhao, Junhui, et al. "A review of active management for distribution networks: current status and future development trends." *Electric Power Components and Systems* 42.3-4 (2014): 280-293.

[2] Niknam, Taher, and Abdollah Kavousi-Fard. "Optimal energy management of smart renewable micro-grids in the reconfigurable systems using adaptive harmony search algorithm." *International Journal of Bio-Inspired Computation* 8.3 (2016): 184-194.

[3] Gazijahani, Farhad Samadi, and Javad Salehi. "Stochastic multi-objective framework for optimal dynamic planning of interconnected microgrids." *IET Renewable Power Generation* 11.14 (2017): 1749-1759.

[4] Wang, Zhaoyu, et al. "Robust optimization based optimal DG placement in microgrids." *IEEE Transactions on Smart Grid* 5.5 (2014): 2173-2182.

[5] Khorram-Nia, Reza, Bahman Bahmani-Firouzi, and Mohsen Simab. "Optimal switching in reconfigurable microgrids considering electric vehicles and renewable energy sources." *Journal of Renewable and Sustainable Energy* 10.4 (2018): 045905.

[6] Olivares, Daniel E., et al. "Trends in microgrid control." *IEEE Transactions on smart grid* 5.4 (2014): 1905-1919.

[7] Wu, Xiong, Xiuli Wang, and Chong Qu. "A hierarchical framework for generation scheduling of microgrids." *IEEE Transactions on Power Delivery* 29.6 (2014): 2448-2457.

[8] Palizban, Omid, Kimmo Kauhaniemi, and Josep M. Guerrero. "Microgrids in active network management—Part I: Hierarchical control, energy storage, virtual power plants, and market participation." *Renewable and Sustainable Energy Reviews* 36 (2014): 428-439.

[9] Bidram, Ali, and Ali Davoudi. "Hierarchical structure of microgrids control system." *IEEE Transactions on Smart Grid* 3.4 (2012): 1963-1976.

[10] Guerrero, Josep M., et al. "Hierarchical control of droop-controlled AC and DC microgrids—A general approach toward standardization." *IEEE Transactions on industrial electronics* 58.1 (2010): 158-172.

[11] Abd El-salam, Mirna, EmanBeshr, and Magdy Eteiba. "A new hybrid technique for minimizing power losses in a distribution system by optimal sizing and siting of distributed generators with network reconfiguration." *Energies* 11.12 (2018): 3351.

[12] Badran, Ola, et al. "Optimal reconfiguration of distribution system connected with distributed generations: A review of different methodologies." *Renewable and Sustainable Energy Reviews* 73 (2017): 854-867.

[13] Jabbari-Sabet, Reza, Seyed-Masoud Moghaddas-Tafreshi, and Seyed-Sattar Mirhoseini. "Microgrid operation and management using probabilistic reconfiguration and unit commitment." *International Journal of Electrical Power & Energy Systems* 75 (2016): 328-336.

[14] Sedighizadeh, Mostafa, et al. "Optimal distribution feeder reconfiguration and generation scheduling for microgrid day-ahead operation in the presence of electric vehicles considering uncertainties." *Journal of Energy Storage* 21 (2019): 58-71.

[15] Huang, YC. "Enhanced-genetic-algorithm-based fuzzy multi-objective approach to distribution network reconfiguration." *IEE Proceedings-Generation, Transmission and Distribution* 149.5 (2002): 615-620.

[16] Li, Yuanzheng, et al. "Multi-objective optimal dispatch of microgrid under uncertainties via interval optimization." *IEEE Transactions on Smart Grid* 10.2 (2017): 2046-2058.

[17] Li, Peng, et al. "Stochastic optimal operation of microgrid based on chaotic binary particle swarm optimization." *IEEE Transactions on Smart Grid* 7.1 (2015): 66-73.

[18] Niknam, Taher, Rasoul Azizipanah-Abarghooee, and Mohammad Rasoul Narimani. "An efficient scenario-based stochastic

programming framework for multi-objective optimal micro-grid operation." *Applied Energy* 99 (2012): 455-470.

[19] Gazijahani, Farhad Samadi, and Javad Salehi. "Robust design of microgrids with reconfigurable topology under severe uncertainty." *IEEE Transactions on Sustainable Energy* 9.2 (2017): 559-569.

[20] Kavousi-Fard, Abdollah, and Amin Khodaei. "Efficient integration of plug-in electric vehicles via reconfigurable microgrids." *Energy* 111 (2016): 653-663.

[21] Dall'Anese, Emiliano, and Georgios B. Giannakis. "Risk-constrained microgrid reconfiguration using group sparsity." *IEEE Transactions on Sustainable Energy* 5.4 (2014): 1415-1425.

[22] Nikmehr, Nima, Sajad Najafi-Ravadanegh, and Amin Khodaei. "Probabilistic optimal scheduling of networked microgrids considering time-based demand response programs under uncertainty." *Applied energy* 198 (2017): 267-279.

[23] Golshannavaz, Sajjad, Saeed Afsharnia, and Pierluigi Siano. "A comprehensive stochastic energy management system in reconfigurable microgrids." *International Journal of Energy Research* 40.11 (2016): 1518-1531.

[24] Hemmati, M., et al. "Risk-based optimal scheduling of reconfigurable smart renewable energy based microgrids." *International Journal of Electrical Power & Energy Systems* 101 (2018): 415-428.

[25] Bornapour, Mosayeb, et al. "Optimal stochastic scheduling of CHP-PEMFC, WT, PV units and hydrogen storage in reconfigurable micro grids considering reliability enhancement." *Energy conversion and management* 150 (2017): 725-741.

[26] Maulik, Avirup, and Debapriya Das. "Optimal operation of a droop-controlled DCMG with generation and load uncertainties." *IET Generation, Transmission & Distribution* 12.12 (2018): 2905-2917.

[27] Kanwar, Neeraj, et al. "Optimal distributed resource planning for microgrids under uncertain environment." *IET Renewable Power Generation* 12.2 (2017): 244-251.

[28] Nafisi, Hamed, et al. "Optimal daily scheduling of reconfiguration based on minimisation of the cost of energy losses and switching operations in microgrids." *IET Generation, Transmission & Distribution* 9.6 (2015): 513-522.

[29] Gazijahani, Farhad Samadi, Sajad Najafi Ravadanegh, and Javad Salehi. "Stochastic multi-objective model for optimal energy exchange optimization of networked microgrids with presence of renewable generation under risk-based strategies." *ISA transactions* 73 (2018): 100-111.

[30] Gaing, Zwe-Lee. "Particle swarm optimization to solving the economic dispatch considering the generator constraints." *IEEE transactions on power systems* 18.3 (2003): 1187-1195.

[31] Chen, Changsong, et al. "Smart energy management system for optimal microgrid economic operation." *IET renewable power generation* 5.3 (2011): 258-267.

[32] Mohammadi, Sirus, Soodabeh Soleymani, and Babak Mozafari. "Scenario-based stochastic operation management of microgrid including wind, photovoltaic, micro-turbine, fuel cell and energy storage devices." *International Journal of Electrical Power & Energy Systems* 54 (2014): 525-535.

[33] Jithendranath, J., and Debapriya Das. "Scenario-based multi-objective optimisation with loadability in islanded microgrids considering load and renewable generation uncertainties." *IET Renewable Power Generation* 13.5 (2019): 785-800.

[34] Gabbar, Hossam A., and Aboelsood Zidan. "Optimal scheduling of interconnected micro energy grids with multiple fuel options." *Sustainable Energy, Grids and Networks* 7 (2016): 80-89.

[35] https://ourworldindata.org/co2-and-other-greenhouse-gas-emissions.

[36] https://www.irena.org/Statistics/View-Data-by-Topic/Costs.

[37] Barker, Philip P., and Robert W. De Mello. "Determining the impact of distributed generation on power systems. I. Radial distribution systems." *2000 Power Engineering Society* Summer Meeting (Cat. No. 00CH37134). Vol. 3. IEEE, 2000.

[38] Chen, Jian, et al. "Optimal sizing for grid-tied microgrids with consideration of joint optimization of planning and operation." *IEEE Transactions on Sustainable Energy* 9.1 (2017): 237-248.

[39] Mariaraja, P., Manigandan, T., & Thiruvenkadam, S "An expert system for distribution system reconfiguration through fuzzy logic and flower pollination algorithm." *Measurement and Control,* vol. 51, no.10, pp.371-382. Nov.2018.

[40] S. Mirjalili, "Moth-Flame Optimization Algorithm: A Novel Natureinspired Heuristic Paradigm," *Knowledge-Based Systems,* Vol.89, pp. 228-249, 2015.

ABOUT THE AUTHORS

S. Thiruvenkadam was born in India. He received a BE degree in Electrical and Electronics Engineering from Bharathiar University in 1999, an ME degree in Power Systems from Annamalai University in 2004, and PhD in Power Distribution System reconfiguration at Anna University, Coimbatore, India. Currently, he is working as a Professor in the Department of Electrical and Electronics Engineering at P. A. College of Engineering and Technology, Tamilnadu, India. He is currently a visiting scholar at Kunsan National University, Kunsan, South Korea. He is a member of the IEEE and Life Member of ISTE. His research interests include software framework, power distribution systems, digital control techniques for power electronic circuits, and power distribution systems.

Ravi Sankar received the PhD degree in Electrical Engineering from the Pennsylvania State University, USA in 1985. He has been with the faculty of Department of Electrical Engineering, the University of South Florida, His research interests include Wireless Communications and Sensors Networking (WPAN/WLAN/WMAN), Speech Processing and Recognition, Applications of Computational Intelligence (Neural Networks and Fuzzy Logic) and Simulation, Modeling, and Design of Intelligent, High Performance and Robust Systems

In-Ho Ra received a PhD degree in Computer Engineering from Chung-Ang University, Seoul, South Korea, in 1995. He has been with the faculty of the School of Computer, Information and Communication Engineering, Kunsan National University, where he is currently a Professor. He was a visiting scholar from Feb. 2007 to Aug. 2008 at the University of South Florida, Tampa, Florida. His major research interests include wireless ad hoc and sensor networks, blockchain, IoT, PS-LTE, and microgrid.

In: Applied Soft Computing Techniques … ISBN: 978-1-53618-180-7
Editors: Amit Kumar Thakur et al.

Chapter 2

BIOMASS

P. S. Ranjit[1,*] and Amit Kumar Thakur[2]
[1]Department of Mechanical Engineering,
Aditya Engineering College (Autonomous),
Surampalem, Andhra Pradesh, India
[2]Department of Aerospace, School of Mechanical Engineering,
Lovely Professional University, Phagwara, Punjab, India

ABSTRACT

In countries including Sudan, Rwanda and Tanzania, Biomass contributes some 33% to the structural-enhancing scrutiny, from roughly 90% to 45% in Pakistan, to 30% in China and Brazil, and to 10-15% in Mexico, which is continually South Africa. These rates also alter irregularly as nations utilize auxiliary companies/oil suppliers daily. The main considerations are that in the next century there will be a reduction in the size of two or more billion citizens currently subject to biomass and what is the method for making the change in the circumstances of this reliance on biomass (neighbourhood and across the world)? "Consistent" [1]. In 1996 the World Bank conceived that' vitality agreements would include the production and usage of biofuels as well as their existing

* Corresponding Author's Email: psranjit1234@gmail.com.

forces.' Different countries draw much from their essential energy from biomass. In essence, 4% of the United States, 14% of Hungary, 18% of Sweden and 20% of Finland. 2 EJ/year, in any event, potentially supplies biomass essentiality, in West Europe and is roughly 4 per cent of the necessary imperatively (54 EJ). The estimate indicates natural capacities between 9.0–13.5 EJ in Europe focused on locations, rates and regenerative shops in 2050.

The biomass demand ranges between 17 and 30 per cent of hard and accelerated criteria expected for 2050. The possibility to modernize biomass stimulation is sufficient to include supporting services, including gasoline and charging of gases and transport fillers, while also adapting to typical applications in biomass [1, 2]. The production of biomass is distinguished from unsustainable energy sources in various natural and socially advantageous environments when spread efficiently and sensitively. This consolidates developed land; work is being conducted, farm surpluses are being utilized in developing nations, the new transport needs to national production networks, pollution reduction, waste disposal and increased reuse are being designed. In comparison with annual arable harvests, the continuous and woody active output is progressively vital for the implementation of alternative stimulating feed. Farmers and the process may deliver numerous beneficial circumstances, except as necessary, in Agro-forestry systems. In order to improve CO_2 absorption, use biomass instead of ineffective power sources to sequester carbon in pine-forested areas will be dramatically advantageous from a monetary perspective [2].

Keywords: biomass, bioenergy, carbon markets, new technologies

INTRODUCTION

One-fifth of the world's productivity comprises of renewable materials, 13%-14% of biomass and 6% of hydro. It is around 25 million barrels of oil a day (55 EJ/year) because of biomass. This is the most influential pillar of the resilience of nation-building (33% of the overall population) for 75% of the world. Due to the inability of individual nations to obtain knowledge on biomass studies and the failure to keep up and update existing datasets periodically, exact biomass standard is not available in the world. Biomass provides 90% or total fertility in certain developed countries. In some industrial countries, for example, the USA (4 for each penny, equal in essential material to 1,9 million barrels of oil per day), Austria (14 per penny), and Sweden (18 per penny) are also used for fertility. Biomass is used in some industrial countries. Biomass is often known and mistakenly treated as a commodity of low quality, but once in a while, when the truth is said, it can be deemed an infinite one equal to petroleum equivalents. It offers fuel adequacy because of its variety and different power sources that can be generated by lawfully chanting for intensity or electricity, or by utilizing extraction technologies, such as aging for alcohol production, biogas manufacturing, and gasification, it is processed into solid, vaporous and fluid fillers. Timberland buildups will also provide biomass for mechanical and rural crops as well as for productivity plants such as trees and sugar sticks to improve their potency directly [2].

Biomass usage has fluctuated widely throughout history and has been profoundly affected by two basic elements: the scale of the population and access to properties. Since annual biomass production, many times as high as the overall usage of global biodiversity may also be naturally generated and used. Despite the release of net CO_2, it would not be controversial to properly discuss this possible cause of vitality dissipation in any discussion. The assumption that by 2050 nearly 90% of the world population lives in developed countries most definitely ensures we will not have so much carbon abundance even if the feasibility of the global trade proposals varies unexpectedly [1].

Also, for the reasons for theoretical hypotheses, not to mention the establishment of structures and the usage of strategies, details on biomass production and use implementation are insufficient. There has been wide-ranging debate in the 1980s on how there is a difference between the supplies of wood from the backwood and country and the usage of fuelwood in different countries. Will Rwanda or Tanzania, for naming two versions, disappear the latter tree by 1987 or 1990 as numerous heroes determine? In these countries, there are always multiple trees! Notwithstanding, we have no thought on whether the first biomass supply and usage estimates were not accurate. We may not know whether to collect knowledge more effectively because there is little cooperation with the open subsidy to improve questions regarding the existence of biomass and the assistance applicable to the production and usage of business fuels.

One of the problems in the emphasis on 'fuelwood' or 'development' as a source of biomass is its neglect of charcoal and the large and frequent usage of harvesting reserves, timber production and compost in different countries—these deposits that represent up to 30–40 percent of the entire supply of biomass. Fuelwood is therefore not completely cut trees, but consisting mostly of twigs and tiny leaves, obtained without substantial injury to plants. Consequently, all biomass systems found over seasons on an annual basis must be incorporated into biomass supply and usage interventions. The usage of this knowledge in the interpretation of trends and in the planning of potential events will be subject to a term scheme. In the standard situation, reports on the utilization of biomass will be supplemented by similar biomass statistics, so that regional and local changes can be tested to assess the depletion of reserves and the conceivability of a supply of biomass. Another relevant issue is the step-by-step guide to help citizens conform to limited biomass sources to ensure safety.

Normally biomass is a method of preparing and gathering square meters, equipment, solids, fertilizers etc., typically not used as an important source not only by nuclear societies or institutions but often by preparing and processing them. Specifically, in cities and industrial environments, this non-cooking usage may be unique. Moreover, the commonplace industrial and small-scale organisations are typically powered by biomass and anticipate critical operation also to become a focal point for existing changes in the natural and domestic ecosystems. Quite some time, biomass imperative is likely to remain the remarkable source of warmth necessary in such organizations as it is currently the primary medium to immense, monetarily sensible and potentially sensible, heat fuel. For instance, these organizations in India have a share of 50% of the gathering area, which, in addition to the agricultural region, offer a considerable amount of public service.

It is difficult to determine the usage of biomass basics, especially in its daily framework, as such, causing more problems. There are two big goals here:

1. Also from times to times will biomass join valid bits of information and when it does so, in general, can it be minimal. Biomass is a modest called fuel. Inaccurate findings from the traditional research in bioenergy such as fuelwood, carbon dioxide, manure and yield advances are given to the problems in deforestation and desertification. In Central Zambia, for example, in the country's unique energy transport region, there has been no indicator of emissions in the world as a consequence of deforestation by consuming wood goods, whether for grain or for coal;
2. Problems with biomass estimation, measurement and management as the dissipated source of imperative or inefficiency of operation have little inherent interest. For starters, charcoal is a notable fuel in several countries that still manufacture significant quantities of alcohol, for starters in Zambia, some 12 percent, and on a dry weight basis about 8–10 percent and Rwanda. For starters, the best broilers in Brazil have productivity of about 35 percent. There are massive possibilities for the competitiveness of charcoal production. Biomass's modestly bad appearance shifts by and large for three key factors:
   - Significant late-year attempts to increasingly develop and realistic photos of existing usage and biomass potential by means of recent tests, reveals, moreover, pilot plants;
   - expanded utilization of biomass, particularly in developing countries, as leading transporter of vitality;
   - Increasing awareness, in addition, of the critical steps for managing net CO_2 emanations, of close and regional, the natural focal points of biomass. In comparison to the general outlook, the utilization of biomass remains unchanging or grows primarily for two reasons: 1) population development;

2). As standards for everyday comforts increase, many citizens in the regional and urban territory in the event of nations turn themselves into diverse biomass uses such as charcoal, wood, construction materials, bungalows etc. In such ways, urbanization will not contribute to a full transition to coal fills. Although the usage of biomass life is a domestic use in the rural areas of developed nations, it is slowly being realized that the urban poor and some small to medium-sized businesses do have significant fuel hotspots. We need to deal with the problems of value for the vulnerable, and particularly for women whose, cooking exercises are an essential factor in biomass existence. Secondly, obviously appropriate usage of the land that needs a fair supply of biomass in all of its feed, electricity, and so on, and thirdly, development and growing standards of regular comforts that involve an increased arrangement of existence. It is mainly in these ways that we work on the efficient production and usage of biomass vitality with the intent that, given the more usual and enormous task, we will offer current forces, for example, power and fluid energy (as is occurring in many rapidly industrial countries across the World). Secondly, as population growth is now greater in the formation of countries, the value in biomass will rise significantly later on. Thirdly, inventions through the development of a fresh or enriched biomass capacity or an enhanced conversion of biofuels through increasingly active vitality carriers, and hence an animation of the value in the foodstuffs, leading to more prominent applications of mechanical environments, primarily by natural contemplation.

The following studies will cover soil accessibility (short to long term), productivities, species and collaborations, ecological sustainability, social aspects, financial activities, subordinate benefits and burdens and hearing issues in the sense of massive bioenergy ventures, irrespective of the size of the ventures in the world or its surroundings. The concerns concerned new processing, manure and fertilizer preconditions, sustainability input/inputs, ecosystem and scene consequences, possible disintegration commitments, future clashes with high output food production lands, and help requirements. A great many of these problems decrease if biomass vitality gives specific advantages in near proximity and worldwide and offers management a long-distance market door to better ground, providing ideal productivities that use less of passages while offering environmental and social benefits [2].

BIOBASED FEEDSTOCK AVAILABILITY AND ISSUES

All-inclusive, plentiful sources of renewable feedstocks are essential for the production of green and profitable biobased industries such as horticultural outputs, accumulation, and ranger-service products. In addition to being included in other systems, including horticultural harvests and planning for accretions, the ranger services properties, dedicated vitality harvests, such as miscanthus and switchgrass, and the civil, heavy waste disposal, lignocellulose biomass is spread worldwide. Within the United States alone, the horticultural and protecting stocks of 1.3 trillion tons are projected at each age. This volume includes extra elements that provide objective details from a number of perspectives on the accessibility of lignocellulosic biomass.

Remember that the concept of biomass does not suggest that it can be obtained monetarily or spontaneously. Of starters, several main U.S. drive crop districts will produce enormous yields from permanent bioenergy crops. The budgetary factors actually do not, though, support substitution. Only attention should be extended to the accessibility of agricultural seed deposits. Moorings use fairly reusable materials, such as maize and wheat, after essential harvests. This exposure to crops in the U.S. Department of Agriculture (USDA) and the DOE ' Billon-Ton details' Stover at the grain proportions has systematically been established based on 1:1 maize stover [3]. Corn stover is considered to be a prime bioprocessing choice, but true indoor accessibility and ranchers' opportunities can not be seen as sufficient within a functioning farm system. However, the Billion-Ton study acknowledges regional (and worldwide) variables impacting the production of crop supplies. For eg, the maize stover, the stalks and accumulations left after grain harvest do not deteriorate quickly in the north atmosphere because of the winter temperatures of the virus. This also leads to difficulties, because the next spring has a lot of stock for field preparation exercises. When biotechnology improves the yields of maize, this problem increasing increase and stove elimination is necessary. This is not the case in the south, as stovetops easily corrupt during rainy and typically warm winters, the rancher is also a major source of the fundamental dirt problem. Rice straw is another easily open-source of yield. Throughout this case, it will have an ecological advantage to remove the straw as, throughout order for a strategic buffer from the rice crop of next year, it must either be eaten or simply discarded next year. For air quality and ranchers, it will be an enormous benefit to create a business opportunity in a biomass framework for this straw. There was a lot of research in California with little success in attempting to develop rice straw markets in reaction to the 1991 Rice Straw Burning Diminution Act. The companies involved developing rice straw production products and bundling supplies. Unfortunately, the stroke includes high silica, which is detrimental to current equipment set and renders developing a comprehensive collecting system impractical. Whatever the case, financial aspects can help the development of increasingly efficient collecting framing if elective innovation may eliminate higher confidence in silica.

The use of wheat straw in biomass handling and biomaterials is feasible since the grain has already been collected in the United States and in other applications [4].

Throughout the near term, maize cobs may refer to the most transparent crop production for the preparation of early legionellosis. The processing of maize cobs in one-pass is feasible and is being developed in the maize ethanol industry in the United States. Demand for maize cobs in some areas of about $80.00 per ton is currently available to be used for the processing of furfural compounds, for instance [5].

Another potential hotspot for lignocellulosic biomass is devoted to vitality crops, the two perennials what's more, annuals. Perennials incorporate yields, for example, miscanthus and switchgrass which have as of late been the focal point of consideration by crop biotechnology and cellulosic ethanol organizations. Perennials offer alternatives to ranchers and land proprietors for utilization of negligible land that is right now in the field or other use. Such contain sequester carbon in their root systems, even as limited amounts of data sources such as manures and pesticides are commonly used. However, the development could later on or as part of a scheme to use negligibly, inadequate ranch property could be a piece of home-grown crop expansion. The financial problems of the distribution of these yields may not match in with replanting column crops. Every year, sweet sorghum and rummage sorghum are included, all of which need marginal sources of knowledge and yield essential biomass. Due to sweet sorghum, a big sugar content may be effectively adjusted with current technologies as ethanol or other biobased goods whereas bagasse may be used as a feedstock for advancing lignocellulosic improvements. Markets are created for this yield in biopower applications, while certain higher yields are popularised. In Europe and in some parts of the United States, there is a growing market field for densified wood and life crops pellets and briquets for house heat, domestic usage and coal co-termination [4]. The above is gradually pushed into the infinite power era and ozone deteriorating hazardous compounds by the organization of administrative needs. In general, referred to as "woody biomass", such materials also benefit from a current year-round system for selection and distribution.

THE JOB OF BIOMASS-BASED ENERGY IN CARBON MARKETS

Economic Significance of Bioenergy

External systems, most of which are the source of oil, are gradually exposed to resilience. The vital transport segment, in particular, is highly reliant on oil supplies. In any case, it is unreasonable to assume, since it is a stable and non-renewable supply base, that oil production stores can sincerely fulfil the increasing demands for viability. Because the current consumption rate which turned away from innovations in technologies and technology was a life cycle of the 1970s versions, oil retailers were indeed forecast to last for around 50 years. Even national security worries the strong energy capacity of the vitality industry, since imports reach the overwhelming majority of US unrefined oil production (57 percent in 2008) (EIA, 2010). Empirical data is often related to a wide variety of scientific questions regarding the usage of non-renewable energy. Such problems have allowed the substitution of biomass, wind, sole and geothermal resources. The more significant part of the above was currently shown to be considerably higher in terms of durability than common stocks, to provide a potential buffer against future monetary instability once the oil asset was depleted. Such modern vitality outlets guarantee sustainability, successful carbon dioxide absorption and innovative finance in the field.

The US Congress aims to achieve 20% of the world's requirements for inexhaustible biomass-based energy resiliency by 2020 and 25% by 2025. by 2020. Worrying over the supply of future oil supplies and environmental risks has helped locate clean fills from infinite US lands. Thanks to a decline in output dependency on imports and other inventory vulnerabilities, ethanol from maize and biomass have collected currency. Biomass is processed into ethanol by conventional maturation, compound refining and finishing. A single maize bushel (56 lbs) will contain ethanol of approximately 2.8 gallons (10 lbs). This material, which contains 35% oxygen, is water-soluble and biodegradable and reduces the reactive chemical compounds used as a liquid at this level. The United States has produced ethanol more than 1000 times as high annually in the past 30 years, within 100 offices around the nation varying from around ten million gallons in 1979 to 10.7 billion gallons in 2009.

Ethanol is also blended with unleaded fuel at fluctuating rates. In the light of development, the prices of ethanol in the United States are already higher, while petrol rates are anticipated to plunge below the dawn of the second innovation era. In countries where ethanol is extracted from various crops, for example, sugar cane, ethanol prices are now well under average gas prices. Ethanol, therefore, has a 33 percent decrease in fuel life and therefore rising the gas miles, thereby gradually viewing costs as financially relevant. Biodiesel is also a medium of elective biomass, which, while the primary source is soybeans, can be made of just about any oil or animal fat. Synthetic biodiesel cosmetics are petrol and gasoline which can be mounted with minimal changes on modern diesel engines. Biodiesel processing awareness costs for plant biomass are low.

Feed waste biodiesel and residual cooking oils offer an efficiency boost while reducing production and depletion of GHGs. The company's office in Arkansas is focused on the biodiesel sector, and Wal-Mart is planning to provide a biodiesel armada of vehicles. Tyson Foods is also a collaborator with Conoco in the development of poultry-fat biodiesel, the world's biggest poultry producer located in Arkansas. The fourth IPCC assessment study forecasts that bioenergy should be offset by 2030, with a stronger estimate of biomass in 2030, a measuring cap, markets and the marketing of second-aged biofuel production. Currently, as well as ethanol from sugar cane in low pay countries, the cost and implementation in respect of CO_2 emissions have remained strategic gap is worrying. The proportion of biofuels used for transport, for example, could increase to about 10 per cent for 2030 by depending on capacity production and development of cutting-edge methods, such as cellulose transformation by the enzyme or by gasification and amalgamation, on the other side.

New Technologies for Present Day Biomass Energy

For certain nations, biomass is an essential position for fertility. However, its current application, including inefficient usage of biomass and poor efficiencies in construction, has a restricted future. However, the growth of collective focus on environment protection and the effective utilization of natural resources are the principal drivers of fundamental improvements in the field of vitality and rivalries between vitality organisations and sources of existence. In this regard, potential biomass options point to current carriers of vitality: strength, fluid and vapour substitution to manage the use of robust feed material. The existing usage of biomass must depend unambiguously on the practicality and financial features of new ways of transformation, or also on the deterioration and preservation of traditional specialization obstacles.

Gasification for Warmth and Power, Methanol and Hydrogen

There is a discussion in this field of warm bio gasification. In relation to cases in which gasifiers are underway, gasification requirements are shown. The applications studied involved heating and power age and amalgamation of methanol and hydrogen gas output. Today's technologies are mainly common for warm applications and power generation. Advancement of business advances for concoction combination is a lot further into what's to come. By gassing biomass, it can be transformed into a steady vapour fuel halfway into an extremely heterogeneous substance that can be efficiently utilized for cooking, mechanical manufacturing, power age and fluid power production. Biomass can be exceptionally healthy wood chips 50 percent wet, dried, with ferrous and other products, industrial forest build-ups, slims and the normal component of the heavy municipal solid waste (MSW). With regard to these materials, the gasification cycle will turn into carbon-and hydrogen-rich fuel gasses and can be utilized much more reliably, and with improved skill and natural execution to manage the combustion of burning biomass. The frameworks include, in general the biomass fuel handling and support system, combined with methods for containing the gasifier area. The gasifier is usually a full filled tank that is expressed with climatic or elevated pressure at temperatures approximately 850^0C.

In order to coordinate the operation, the aspect of gas has to be controlled. With closely-knit gasifier-combustor setups, gases are not washed. To order to insure that a turbine heating area is appropriate in the gas turbine applications in the context of regulation, the gas shall be clean of dub, tar, arsenic, chlorine and salt metals. In internal burning applications, the gas must be cooled in order to ensure that a sufficient life charges can be put into each chamber and to insure the valves and chambers are protected with special regard to tea and particulate matter [4]. For applications requiring energy materials, gas will often need hydrogen to guarantee that the cathodes will not have some significant sulfur or chloride emissions. To order to avoid downstream damage, particulates and toxins (e.g., H_2S) must be removed for amalgamating operations, for instance, to produce methanol and hydrogen.

Process output is very high: for each penny, warm gasification usually requires 80–85 to transform the feed's natural material into a gas mixture—the usage of linked cycles to produce electricity tests intense biomass for power efficiencies above 45 percent. The decreased efficiency of power sources will be more than 55 per cent as the gasses are converted to hydrogen (HHV premise). The beneficial, environmentally friendly condition is that fuel gas is far smaller than the evaporator burning current. Typhoon as salts may be discarded owing to this, and the mostly lower treatment temperature of the biomass allows the preservation of metals (counting salts) in the waste. The gas will be extracted before combustion without any of the magnitudes of the severity of corrosive product fragments like hydrogen chloride.

Gasification

The structure of the gasifications-based cycle depends, however, on the form of the reactor.

Gasifier Medium

Gasification entails oxidation and biomass processing in steam air or air (or both) for the medium to low heat gas supply [6]. The exothermic reaction of oxygen and organic matter is used by air-or lawfully heated gasses for deforestation of vegetation to change the heat in the remains of carbon-rust. In such legally warmed gasifiers, the heating to drive the cycle is produced inside the gasifier. Once air is used, the resulting gas is strengthened with nitrogen and typically has a heat value of around 5–6 MJ/Nm^3 in the dry premise [4]. By using oxygen rather than air, the dry-premise level for heating agent gas can be raised to 13-14 MJ/Nm^3. Oxygen production is costly, though, and its use is suggested exclusive for direct heating gasification purposes, including amalgamation gas generation, when the use of nitrogen in downstream mixing practices is not allowed. Oxygen usually costs from 40-60 dollars a ton and is used at a scale of 0.25–0.3 tonnes/ton of biomass, at the expense of 10-20 dollars a ton of biomass. [6].

In comparison to the air blown method, warmed gasification is an alternative. Biomass is cooked or gasified through a high-temperature membrane by storing hot gasses. Oxygen is not used such that a medium calorified gas with small nitrogen diluents is provided. Dry premises are typically common for 18-20 MJ/Nm^3 estimates.

Pressure

Another attribute that influences the output of gasification-based power frames is the gasifier working weight. The show and the financial concerns of gasification-based energy frames and gas turbine-based cycles are especially important from this perspective.

Pressure proportions of turbines of 10–20 are common, resulting in a 1.0-2.0 MPa turbine delta. A pressurized gasifier produces gas at a weight appropriate for direct control of the turbine which offers the optimum performance before and after the cycle. To order to play the maximum advantage of working under control, specific organizational systems must be formed. Effective, high-pressure feed systems were not monetized. During mill gasifier conditions (825^{0}C and 2 MPa), tar, monkeys and volatile salts are generated. To order to improve performance without reducing the temperature below the tar dewpoint, these products will typically be separated from the hot product gas at approximately 540^{0}C. The gas systems are smooth-up. Tars normally provide a high degree of heat which can be combusted into fuels but may obstruct roast canals which burning frame ashes that damage the integrity of the burning better and lower flickering ignition process (to avoid radiative heat transfer problems on the turbine) is ensured by the elimination of tars. It may be a synergistic or warm tar wafer that is an important part of a rehabilitating process for hot gas. A synergistic tar wafer typically operates at 870–980^{0}C, at virtually equal gasification temperatures (approximately 825^{0}C) and moist saltine [6].

The gas portion is cooled for the volume of soluble base gases on tar saltine at a minimum range of 350-650^{0}C. The target at this stage goes into a fired pipe, which often calls for the removal of a significant portion of the water. A unique displacement is planned to avoid rigid turbine tops from disintegrating. In any event, fuel may generate an inappropriate amount of soluble foundation for gas turbine applications, because the gas turbine cap of antacid is around 25 ppb in the turbine gas escape. A porous base bed may be added after the pipeline, thus removing affidavit and utilizing turbine-cutting edge content, to guarantee that the gas complies with solutions-soluble specific requirements. The basic bed materials are malachite or hectorite. Getters would presumably be planed as parallel fixed sheets, taking bed replacement into account the continuous movement.

It is also necessary to handle the gasifier with low weight and to stack the filtered liquids to the weight needed to use for a gas turbine. In this case, tar saltine is more commonly used to restrict the amount after the disappearance of tar. The tar saltine's gas supply is equilibrated, so the blower supply is decent. The water and tar would be reasonably low to ensure no change in loads. Likewise, solvent tars which influence the characteristics of grass oil should be dispensed. Therefore, a heating system is used to eliminate gas from the tar to the tar dew point and wet scouring. The smoking substance should be immersed at a scrubber exit temperature (approximately 90^{0}C) and pressure.

Gasifier Type

Three main forms of systems have been built for biomass-generating reactors: Fixed bed, Permeating Fluid bed and Coursing Fluid Bed [4].

Fixed Bed

Updraft and downdraft bed gasifiers may primarily be treated. Within the update, you can consider the most powerful and least effective gasifiers. The new gasifier is a counter-flowing reactor, which supplies fuel with bottle or valve locks techniques and then flows down through the reactor to the waste expulsion feature. The gassing source, air or oxygen and through the reactor possible steam are transmitted upwards. Usually, 80–100^{0}C is caused by temperature. A ton of condensable tars and oils in the pyrolysis area may be used for this reason. In close proximity mode, enhanced gasifiers are used routinely for transporting steam or spraying water through a radiator or tub. Many supports with low-consolidated debris can slag in the expendable cross segments. To order to maintain the feed particle size, a normal bed should also be handled.

In the crisis gasification, the atmosphere and the material exactly suit the firm sheet. Downdraft gasifiers are built explicitly to restrict tar and oil production. Petrol and pyrolytic gasses are now going into the room. The left gas normally has a temperature of about 700^{0}C. Downdraft gasifiers have ambiguous general goals on the feed properties of changed gasifiers. To preserve the physical bed properties and define the perimeter, the feed molecule must be fully uniform with no fines. The feed has to be thin, compact junk with a consistent mixing temperature. The instant food content should also not be 20% to sustain the requisite high temperatures for the isolation of tar. A minor difference from the gasifier is the intersection gasifier in the surface, which is shown digressively on the base of the gasifier. The working state of the crossflow gasifier is not different from that of the downflow gasifier.

Permeating Fluid Bed

The gas flow in a fluidized bed of gas increases to a bed of loose, bubbling granular substances where the gas pressure is huge enough to detach and continuously scatter large particles around the surface. Gas fluctuates in channels of tiny solids and bundles or stacks of solids downstream during the entire cycle of dispersing the pad [7]. The liquidated bed is fluid-like and has solvent-like electrical properties. The solvent consisting of the sand, chalk, dolomite or alumina in the stream-bed gassing cycle of biomass, and is dust, water, or steam. A storm is used either to send fines to bed or to remove garbage from the pipe [8]. Biomass should be shown by a feed box in order to obtain the most significant reward behind a pillow. The incorporation into bed creates breathing space for fines, which will also fluidize the gas and not be changed in the room. The bed is continually pre-heated with an exterior flame, propane or, for instance, fuel type. Fluidized bed gasifiers offer high temperature, an impressive upside downmix with remarkably stable bed conditions. According to this, a character of gasification is 95–99 percent carbon exchange. Mumbling fluidized-bed gasifiers are usually recommended for the full residues that require the development of tropical tempests.

Coursing Fluid Bed

Fuel flowers are visible; the bed is surrounded by immense vacuums, and the fuel movement of a flushing liquid bed contains vast volumes of solids. This sort of bed is an incensed liquid bed. The gross bed solids are put in a bubble bath that is separated from the gas and applied to the soil as a strong river belt. The wide solid region and the hollow solid atmosphere are not easily differentiated, and a circumnavigational liquid bed may be made of a saturated liquid bed. Densities of roughly 560 kg/m^3 in spherical liquid beds are distinctly seen when opposed to an excess volume of 720 kg/m^3 [9]. The gas pace from entering beds is increased from 1.5-3.7 m/s to around 9.1 m/s to hit the thickness of the lower sheet. The length of the solids in a liquid waste bed is dependent on the solid flow rate, lower solids and the capacity to provide the contraction of the solid divide.

Entrained-Flow

The feeding in a pneumatic reactor induced by the induced-stream gasses, dry or slurry, is continually improved with a big oxygen stage. At elevated temperatures, oils and tars absolutely decimate oxygen. The high temperature (from 130^{0}C to 140^{0}C) means, however, that waste is regularly eliminated as a flux. For coal and biomass analysis, these gasification agents were made, without any doubt, minimal. Non-application of biomass has many reasons, but the key issue is the essential expense of feed supply to increased moisture to reduced and decreased molecular size [10].

Methanol and Hydrogen Generation

This section explores the thermochemical production of methanol and biomass hydrogen and addresses financial problems rapidly. In the light of the existing possible tests, the propensity for temporary impact, and the tremendous degree of modern suspense, the gasification for biomass for synergistic methanol or hydrogen amalgamation combinations is the invention selected for analysis and growth. A good talk will not be offered of the reacting science of steam change, water gas movement and methanol fusion. Twigg points to knowledge on the reactant chemistry of these approaches [11]. For a general exchange of combination gas change to fluid energizes to Mills [12].

Methanol Generation Innovation Depiction

The thermochemical processing of biomass methanol involves the development of the H_2-rich mixture gas CO, which is then chemically converted into methanol. Warm gasification triggers amalgamation gas. The methanol-generation unit tasks of biomass are divided into the related important areas: 1) feeding; 2) gasification; 3) gas modification combination; and 4) methanol mixture, refining. The downstream equipment of the gasifier matches that used for the production of petroleum gas methanol. The gasifier syngas is then used for the methanol mixture following proper cleaning, upgrade and redesign. Eventually, the methanol component undergoes the methanol purification process, which expels water and natural materials in varying degrees as the current usage of methanol suggests. Several gasifiers that may generate a union gas fair for amalgamation of methanol are in development in order to limit innovation. These gasifiers operate at a rate of 4-200 mg/d. Both systems are designed to provide small to medium projected fuel power. No structure has been built on a systematic system to decide work parameters which are essential to the development of the largest methanol. All downstream fusion gas operations reflect market development under which operating conditions and yields are established. Recent technologies have scope, for example, the amalgamation of single-pass fluid phase methanol.

Hydrogen Generation

Hydrogen production is an advancement in the petroleum refining sector in which steam reforming invention and decontamination is used to convert light hydrocarbons such as flammable coal, plant fuel coal refined, molten petrol gas/butane and light hydrogen Naphtha, petrochemicals, metallurgical preparations and preparations for feeding, hydrocracking or other manufacturing equipment. For examples, heavier hydrocarbons may also be used to generate hydrogen by means of partial oxidation. [13]. For combination gas change steam improving is the choice considered.

In more established steam-improving units [13], four handling steps are included:

1) vapour modification to CO and H_2 by methane; 2) variable CO to H_2 by transformation; 3) vapour refinement through CO_2 adsorption; and 4) the methanation for the elimination of small CO_2 and CO remaining volumes. Modern vapour enhancement techniques are fairly exceptional in the presence of advancement in industrial pressure vibrating adsorption. Biogas is turned into steam as the weight is raised (usually 3.5 MPa), and the element then moves into reactors with alternating high and low temperatures to extract CO. A PSA device is then invited to supply high-virtue hydrogen to gas from the low-temperature move-in reactor. The PSA's off-gas is used for steam reforming as a heat. Princeton College has produced hydrogen cost gages from biogas [10, 14]. Costs of hydrogen production vary from US$ 8.71/GJ H_2 to US$ 17.46/GJ H_2 to 1,000 Mg/d IGT dependent cycle with a 1.650 Mg/d feeding of the battelle gasifier. For a US$ 6.26/GJ H_2 procedure with 1,120 GJ H_2/d, nearly equivalent hydrogen prices would be paid at US$ 7.90/GJ H_2 for a procedure of 712 GJ H_2/d [15].

REFERENCES

[1] www.fao.org.

[2] Hall D.O. Biomass energy in industrialised countries—a view of the future. *Forest Ecology and Management* 1997;91:17-45.

[3] Perlack R.D., Wright, L. L., Turhollow, A. F., Graham, R. L., Stokes, B. J., Erbach, D. C. *Biomass as Feedstock for a Bioenergy and Bioproducts Industry: The Technical Feasibility of a Billion-Ton Annual Supply,*. 2005.

[4] Riegel's Ka. *Handbook of Industrial Chemistry and Biotechnology.* Springer.

[5] Khanal S.K.L, B. P. *Bioenergy and biofuel from biowastes and biomass* 2010 pp.1-22 ref.28.

[6] Bain R.L., Overend R.P., Craig K.R. Biomass-fired power generation. *Fuel Processing Technology* 1998;54:1-16.

[7] Perry R.Ha.C., C.H. *Chemical Engineers' Handbook.* 5th ed: Mc Graw Hill; 1973.

[8] Hansen J.E. Fluidized bed combustion of biomass: an overview. In: Energy RUDo, editor. *Proceedings of Biomass Combustion Conference*. Washington DC1992.

[9] Wilcox B.A. *Steam.* 40 ed: Barberton, USA; 1992.

[10] Larson E.D., Katofsky R.E. *Production of Hydrogen and Methanol from Biomass.* 1992.

[11] Twigg M.V. *Catalyst Handbook.* 2nd, 1999 ed. London: Wolfe Publishing; 1989.

[12] Mills G. *Status and Future Opportunities for Conversion of Synthesis Gas to Liquid Energy Fuels: Final Report, Center for Catalytic Science and Technology.* 1993.

[13] Gary HHaH, G.E. *Petroleum Refining Technology and Economics.* 2nd ed: Marcel Dekker; 1984.

[14] DeLuchi M.A, Larson, E.D. and Williams, R.H. *Hydrogen and Methanol Production from Biomass and Use in Fuel Cell and Internal Combustion Engine Vehicles: A Preliminary Assessment.* 1991.

[15] Padró CEGaP, V. *Survey of the Economics of Hydrogen Technologies.* 1999.

In: Applied Soft Computing Techniques … ISBN: 978-1-53618-180-7
Editors: Amit Kumar Thakur et al.

Chapter 3

HYDROGEN: PRESENT AND FUTURE ENERGY

P. S. Ranjit*
Department of Mechanical Engineering, Aditya Engineering College (Autonomous), Surampalem, Andhra Pradesh, India

ABSTRACT

Hydrogen may be classified as a renewable energy carrier next to electricity. Hydrogen development includes a range of domestic choices, including green energies and nuclear plants. Over the long term, hydrogen also reduces the dependency on imported energy and greenhouse gas emissions and other pollutants. Hydrogen may be considered the fundamental element in existence. Hydrogen is one of the essential elements of the earth's crust. Hydrogen, as a substance, is not necessarily contained and must be generated on earth. The explanation is that hydrogen gas is lighter and thus sinks into the atmosphere. Relevant components in the form of a composite, including biomass, gas and oil, also contribute to natural hydrogen.

The maximum energy density of hydrogen has that weight gasoline. The lowest carbon content is methane, on the other hand. It's a gas which is the lightest component of it at the average temperature and pressure.

* Corresponding Author's Email: psranjit1234@gmail.com.

Hydrogen is the secondary energy supply, commonly called a fuel tank. Energy carriers are used to move, store and deliver electricity in the simplest way. Electricity is the most common form of power supply.

Hydrogen would have a range of benefits as a theoretically sturdy electricity carrier.For some instances, a considerable amount of hydrogen is healthy to carry, such as deep fuel storage, liquid hydrogen for tanks, and the release of hydrogen by chemical compounds through heating. Additionally, hydrogen is used as an active, low pollutant fuel that is ideal for transportation, heating and power production in places where electricity is hard to find. The distribution of hydrogen through pipes is, in some instances, simpler than for long-range cables.

Hydrogen may also be used to improve the latest IC engine technology in additional mode before being transformed into a small hydrogen generator. Unwanted ignition activity was also tried in tandem with induction strategies.

Keywords: hydrogen, present & future energy, supplementation, IC engines

INTRODUCTION

Hydrogen is and should remain an essential source of energy given its safe combustion and potential availability from renewable sources [1-4]. The critical policy preparation plans for renewable energy ventures are often an alternative for utilizing hydrogen as a power carrier[5]; the Japanese Ministry of Economy, Trade and Industry [6]; the European Commission's Directorate-General for Research [7]; the United States Department of Energy [8] Then some people published discovery trials and then provided funding requests for hydrogen. Hydrogen is primarily found in internal combustion engines, and as an interim form of energy, the conversion is proposed as a non-carbon source until the fuel cell advent [9]. Hydrogen without odours, without odour, without colours, without degradation and non-toxic. Combustion properties further discuss the aforementioned. The following.

Wide Flammability Range: Compared with other liquids, hydrogen has a broad range from 4 to 75% volume inflammability. The hydrogen in the air mix is ideal for stabilizing the fuel mix at air pressure and in the room temperature. The spectrum of 1.6 to 7.6 percent is between 0.6 and 5.5 percent, between 2.15 and 9.6 percent, between 1.9 and 3.7 percent, and between 7 and 14 percent, respectively for oil, fuel, LPG, CNG, ethyl ether (DEE) and biogas [10]. The health of storing and processing depends on massive inflammation. The extreme inflammation of the hydrogen allows mixing of gasoline with a wide variety of fluid leaning from oil to gas or fluid air, where the gasoline element is streaming under the chemical mix [11].

i. Quenching Distance: Hydrogen is raised to 2 mm, with a 0.6 mm gap from the inner cylindrical area, where the fire becomes redundant. Consequently, the present approach will make fires more difficult to quench in comparison to other fossil fuels more vulnerable to hydrogen flames [11].

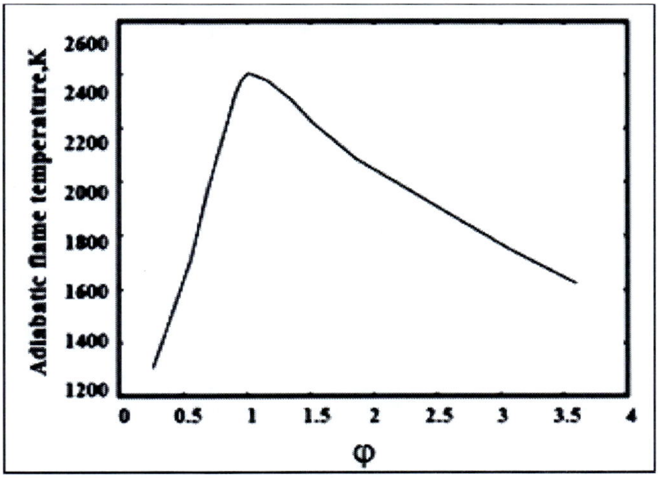

Figure 1. Adiabatic flame temperature for H_2-Air mixture [13-15].

ii. The velocity ofFalme and its Adiabatic characteristics: Hydrogen has an extreme fire rate such that hydrogen engines become more tightly controlled until the stoichiometric air-hydrogen combination has been burned. Instead, fire rate reduces dramatically owing to the fuel-air equilibrium as it runs. The laminar hydrogen flame frequency is 1.9 m high per second at natural pressures and temperatures equivalent to five times the amount of a hydrocarbon liquid.The properties for use in internal combustion engines for hydrogen, particularly thermal performance, combustion strength and contaminant, fire and adiabatic temperatures, are fundamental in Figures 1 and 2 [11-13].

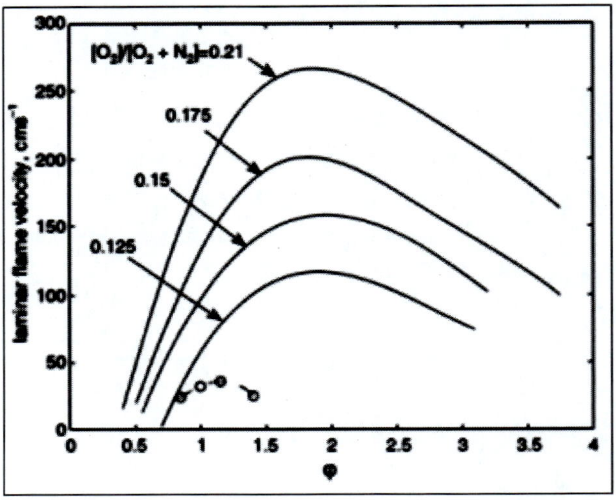

Figure 2. Laminar inflammatory speed for (—), mixtures of hydrogen, oxygen, nitrogen and (always, –) petrol and air[13-15].

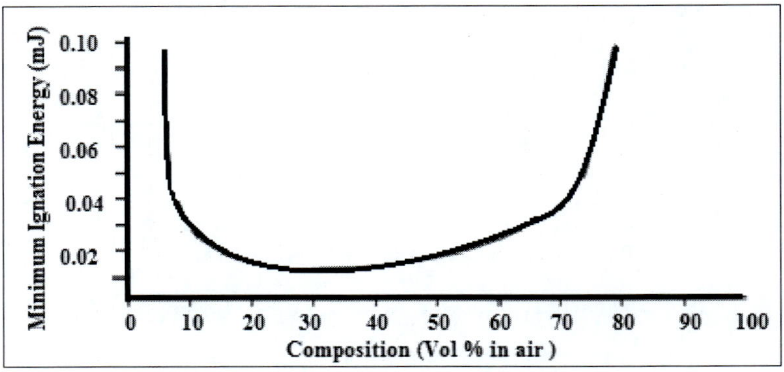

Figure 3. Minimum ignition energy of H_2 –air mixture.

iii. The requirement of Ignition Energy: The ignition source combustion mixture requires limited inflammatory ability. Total hydrogen and air-fluid combustion temperatures are 0.02 mJ, with fuel oil at 0.24 mJ. The low ignition capacity, however, helps the hydrogen blends to interact with accessible hot spots and heat fuels, as shown in Figure 3 to guarantee early start [14].
iv. High diffusivity: In comparison to other gasses, hydrogen is somewhat diffusive. Such a beneficial feature will lead to the formation of the standardized gas mix and thus boost the combustion condition in the CI engines. The air emission efficiency is significantly higher than fossil fuels. Nonetheless, relative to TDC, a homogenous blend is simpler and more accurate. The power is improved to split thermally. To trying to avoid hazardous conditions such as water spills, it would also be very quickly transferred to the community [15].
v. Low Density: The power per unit amount is smaller, owing to the high level of hydrogen. Significant compression and conversion to a hydrogen liquid state are required. Low density also implies a level of fuel-air solution that reduces the energy output of the engine [11].
vi. High Self-Ignition Temperature condition: The temperature of self-ignition is vital in case of autonomous combustion in a flammable liquid without any external source. The allusion to itself is like 585^0C hydrogen. The self-ignition temperature affects dramatically when a hydrogen-air solution is robust. The temperature of self-ignition is an extremely critical consideration when you find the engine compression ratio. This H_2-air mixture needs a substantial degree of compression. Nevertheless, high compression ratios in hydrogen assisted or replacement engines may lead to severe hit [15] and [16].

Table 1. Essential properties of hydrogen and other fuels [2]

Fuel	LCV (MJ/kg)	HCV (MJ/kg)	Stoichiometric Air-fuel ratio (kg)	Combustible range (%)	Flame Temp. (°C)	Min. Ignition Energy (mJ)	Self-ignition temp. (°C)
Methane	50.0	55.5	17.2	5-15	1914	0.30	540-630
Propane	45.6	50.3	15.6	2.1-9.5	1925	0.30	450
Octane	47.9	15.1	0.31	0.95-6	1980	0.26	415
Methanol	18.0	22.7	6.5	6.7-36	1870	0.14	460
Hydrogen	119.9	141.6	34.3	4-75	2207	0.017	585
Gasoline	44.5	47.3	14.6	1.3-7.1	2307	0.29	260-460
Diesel	42.5	44.8	14.5	0.6-5.5	2327	-	180-320

Table 2. Properties of H_2 [4, 17]

Properties	Diesel	Unleaded Gasoline	Hydrogen
Self- Ignition Temp. (K)	530	533 – 733	841-858
Min. Ignition energy (mJ)	-	0.24	0.02
Flammable limit (vol. % in air)	0.7 - 5	1.4 – 7.6	4 -75
The stoichiometric air-fuel ratio on mass	14.5	14.6	34.3
Limits of flammability (equivalence ratio)	-	0.7 – 3.8	0.1-7.1
Density at 16^0 and 1.01 bar	833 – 881	721 – 785	0.0838
Net heating value (MJ/kg)	42.5	43.9	119.93
Flame velocity (cm/s)	30	37-43	265-325
Quenching gap in NTP air (cm)	-	0.2	0.064
Diffusivity in the air (cm^2/s)	-	0.08	0.63
Octane number	-	92-98	130
Cetane number	44-55	-17	-

USE OF HYDROGEN AS A SUPPLEMENTED MODE WITH CONVENTIONAL DIESEL, VEGETABLE OILS AND THEIR BLENDS

Hydrogen has 567^0C, which is far higher than other CI diesel, and a good self-ignition temperature. Researchers need an external ignition source to fire the hydrogen-air high-temperature mixture. Wong [18] and Homan [19] It was proposed that it is difficult without any external ignition source to work on the single working hydrogen compression ignition engines [18-22]. The external ignition mechanism such as spark plug and pilot plug and fossil fuels injectors were suggested and tested successfully. For these, a reliable hydrogen ignitions trigger, called the two-mode action in the compression stroke, was found to be pilot fuel injection.

Nevertheless, the dual-fuel mode tends to be the most typical and more efficient combustion when the conventional diesel hydrogen is applied. The standard of inlet air was usually used with hydrogen from the inlet multiplier. The hydrogen produced is appropriately mixed with air before being injected as pilot oxygen into the compression stroke. The hydrogen-air combination is compressed and reached the optimum temperature to fire hot diesel furnace. Nevertheless, this temperature is inadequate to kill the strong self-ignition potential of the regulated hydrogen. The only factor that leads to the self-ignition of the mixture is atomisation, vaporisation and association of the dry hydrogen-air combining. The pilot fuel provides strength to spark the hydrogen. Furthermore, dual-fuel CI engines have the potential for reduced exhaust emissions and improved engine efficiency [23].

When hydrogen is used in the compression ignition engine intake manifold during suction, the hydrogen burn depends on how similar or mixed the following phase is.

(1) Where the combustion cycle needs large, rich hydrogen mix-air to enable the initiation and propagation of several turbulent flames.
(2) Because of improved compression level, hydrogen and pilot oxygen, combustion chamber inflammation and heat lead to the thermal transfer of hydrogen and air mixtures. Unburnt.

Nevertheless, the effect on the efficiency, emissions and combustion of hydrogen supplementation/substitution was mentioned below in compression ignition engines.

Haragopal Rao B. et al. [24], Their research on the mono, water-cooled, 5 HP, 1500 rpm diesel engine and supplied 10 percent of the engine's total energy supply with hydrogen fuel. We observed that at high loads instead of low loads, the average thermal performance was used. The overall hydrocarbon content in the engine exhaust was not evident at any loads with the addition of the hydrogen into air use because of the weak hydrocarbon fuel provided by hydrogen. By contrast, NOx has been raised owing to the improved thermal escape because of the accelerated release of carbon.At high loads, the largest cycle stress differential was recorded, in comparison to only gasoline, gross cylinder (Pmax), where the load was augmented by hydrogens instead of the 10 per cent increase in hydrogen with fuel-compression allergy. The ignition delay was prominent at maximum load while hydrogen was added. The elevated production of Pmax is induced by intensified combustion. The rate of pressure rise was not significantly impacted.

Li Jing and Lu Ying-Qing and Wang Weing [4, 25, 26], Multiple experiments have been performed at various speeds between 2600 and 3000 rpm with injectors of 21.5^{0}CA and 20^{0}CA respectively. Obtained interesting results in comparison to a loop pure diesel in a dual fuel with hydrogen in high loads. Mathur H B et al. [27], Limited hydrogen supply tests with a four kW, mono tank, water-cooled CI motor with direct hydrogen supplies up to 38 per cent of total fuel consumption have been carried out. Significant cuts were made in smoke pollution, and NOx pollution was decreased. The 50% hydrogen substitute emissions rose by 20%, 30%, 40% and 40 lpm smoke, respectively, to 15.6%, 22.2%, 26.7% and 33.3%, and 100% decreased, respectively, by 16.42%, 24%, 28.4% and 37.4%, respectively.

Naber J D et al. [28]. The study shows the important behaviour and successes of delays in the fuel injection strain of combustion and the output of hydrogen. During hydrogen combustion with clear glass, a specially built disco-shaped combustion chamber was used. The inflammation period was lower, the temperature of the air-gas was over 840 K, and the amount of oxygen was 5 to 21 percent. In terms of atmospheric strain, as the fuel temperature was raised, the influence of ignition delays was also low. It was identified, too. They claimed that the late hydrogen injection process enables CEV pollution requirements to be met without any post-gas constraints.

Senthil Kumar M et al. [1], Extensive testing The4-stroke, single-cylinder, direct fuel oil SVO is conducted with 3.7 kW. The petrol-driven injection is improved by a petrol hydrogen engine inlet manifold.

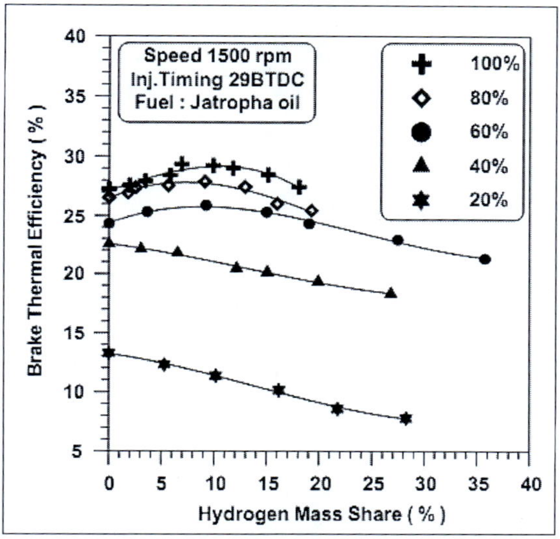

Figure 4. TE vs. hydrogen mass share of a Jatropha based SVO.

They say that the Jatropha straight vegetable oil split thermal capacity improved by 27.3 percent to 29.4 percent over the maximum load with 7 percent hydrogen mass relative to the pure vegetable oil motor. [17]. By comparison, the thermal brake performance of a pilot has dropped from 30% to 32% with a weight of 5% hydrogen. Gases dropped from 4.4 to 3.7 BSU at optimum performance.It rose at full strength from 735 to 875: Max speed, double fuel hydrogen replacement. Prevent the pause in growing oil and fuel crops. The combustion time was shortened because of the high fuel density of hydrogen. Higher pre-mixed hydrogen induction burning levels were recorded [10].

Saravanan N et al.[29] Limited strength, single, four-speed, water-cooled, DI CI, reliability and emissions were concerned. With a hydrogen injector, the injector portion of the valve was mounted in the region of 13 mm. Hydrogen injection begins at 5^0bTDC and continues at 15^0 aTDC with 30^0CA control at the pace of 90^0CA. Figures 5 and 6 display the specific energy usage at maximum load by5^0 aTDC hydrogen timing with a 30^{0}CA injection period. That is because oxygen and hydrogen in the combustion chamber have greater oxidation. At 15^{0}ATDC with an additive of the order of 5^{0}ATDC and a 90^{0}CA injection period compared to 29.4% in the engine knock on the test, thermal brake (BTE), with a maximum of 60^{0}CA, increased from 23.6% to 31.67%. More similar pollution decreased by a higher equivalent of 1806 ppm pure diesel at a pace between 60^0 and 705 ppm [4, 17].

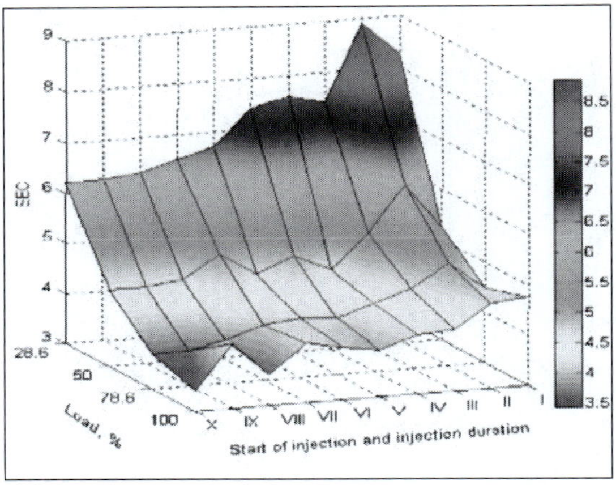

Figure 5. Influence of SEC with load.

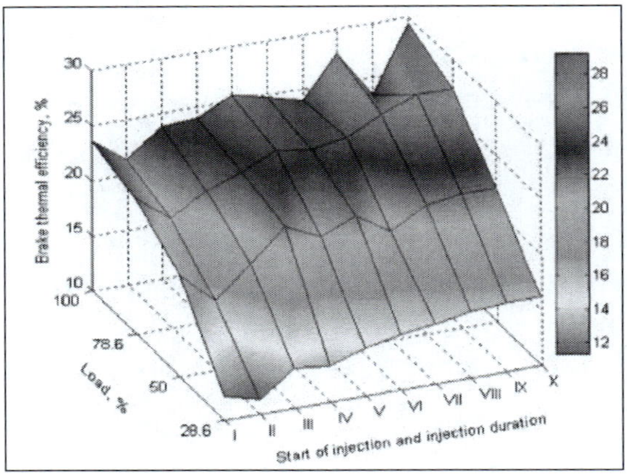

Figure 6. Variation of BTE with load.

Figures 5 and 6 show a reduction in the energy sensitive use of the 5^0aTDC timing at peak load with 30^0CA injection. So oxygen and hydrogen are often oxidized in the combustion chamber. The load at injection begins at 5^0 aTDC with a volume of 90^0 of additional hydrogens and pure diesel operations at 9.5 percent at volume to 2.1 percent by volume to 75 percent. In Figure 10, maximum pressure for hydrogen replacement and pure diesel service was rising from 73.7bar to 71.7 bar with a full load. [10].

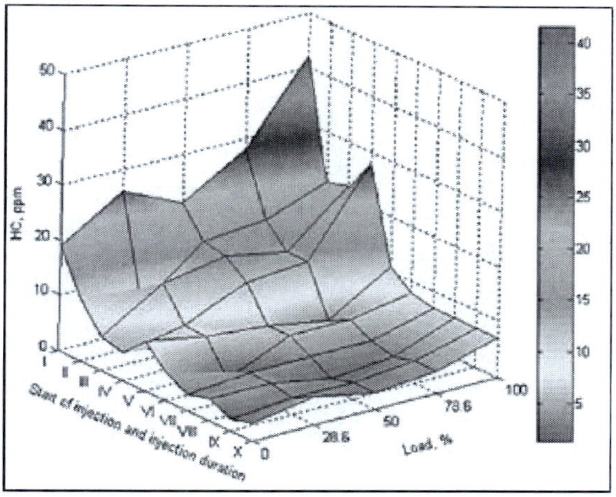

Figure 7. Influence of HC with load.

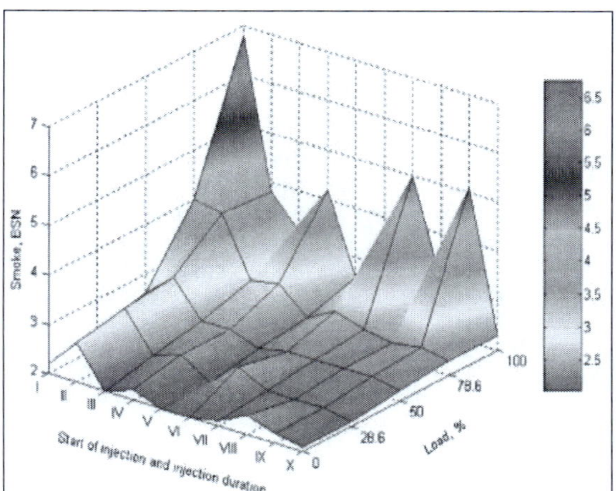

Figure 8. Influence of smoke with load.

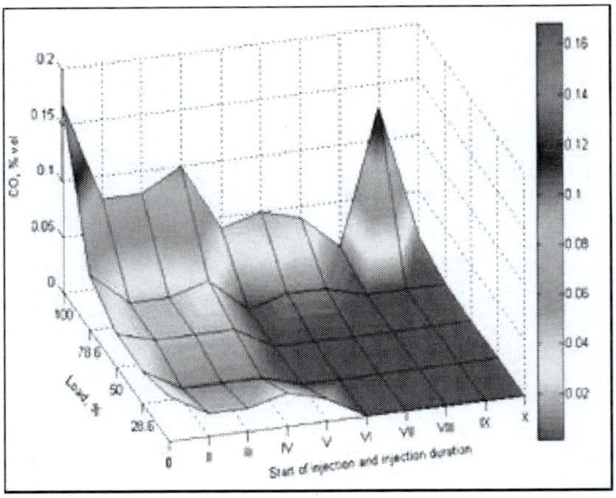

Figure 9. Influence of CO with load.

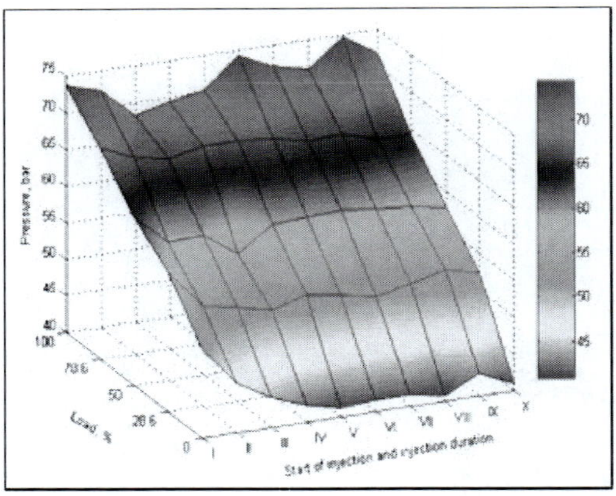

Figure 10. Influence of peak pressure with load.

Edwin Geo V et al. [30], There have been double-fuel diesel engine studies. Fuel for the application of oil (RSO), Rubber seed (Hevea Brasiliensis) methylated ester (RSOME) oils. RSO, RSOME and diesel hydrate's total market share was 8.39%, 8.73% or 10.17%.The typical RSO hydrogen proportion of overall thermal output is 28.12%, RSOME 29.26% and diesel 31.62%. An appropriate RSO and RSOME combination of 8.39% and 8.73% is controlled without any issues with both 10.76% and 11.2%.Later phases of hydrogen enrichment were found for the last knock.

The high loading and loading gain at full loads is improved in relation to the partial load levels [17].

Matthew G Shrik [31] Two independent experiments were conducted by engines. The first series of tests were performed at 1.3 litres, 53 kW, compression Ignition and eddy-current engine dynamometer, and other experiments at 1.3 litres, compression 66 kW ignition engine coupled with a frame and urban moving dynamometer were carried out.All motors are powered by B20 and hydrogen increased by 0-10% for overall power usage of B20, up 5%. The power production was not so influenced with the dynamometer, although the dynamometer of the frame marginally decreased.

Saravanan N et al. [32], Several single-cylinder tests have been performed, including 4 strokes, compression strength, steady velocity, water-cooled, linear, quick, low power diesel and hydrostatic fuel substitution [17]. The efficiency of thermal braking in hydrogen substitution has increased from 22.78% to 27.9% at 30%. Different energy usage (SEC) reduced the amount of hydrogen. For an increased share of 30 per cent hydrogen, rather than 14.5 MJ/kW-hr for pure fuel, the SEC has been lowered to 12.9 MJ/kW-hr for maximum load. Regardless of the drop in the temperature of the combustion, the proportion of NOx hydrogen decreased by up to 30% after sharing. It is 4.6 bar less than pure diesel, and is 6.8 bar greater than pure fuel and has improved to 71.1 bar with maximum load. Lower hydrogen burns generate high peak pressure which leads to decreased cylinder pressure. The process of burning is almost instantaneous.

Same researchers [33] Extended gas exhaust recycling (EGR) monitoring on the same system at different flow speeds. The usage of exhaust gas recirculation is primarily used in two methods. The first is substitution and the second is waste gas recovery. The EGR dump replaced the waste gas the air supply [17]. Furthermore, exhaust gasses are mostly used in compression ignition motors in order to facilitate air intake.Dilution, thermal and chemical impacts of EGR were found to minimize NOx emissions. The very powerful effect of dilution was shown. The best option was found to be 20% EGR in this called dual-fuel hydrogen engine. It 20% EGR named parameters of efficiency and emission.From 15.6 percent to 23.01 percent and 24.2 respectively, the thermal performance division (BTE) rose with and without EGR. The BTE has been marginally diminished by EGR. The EGR amount was 184 ppm to 2212 ppm, with NOx concentrations falling from 99 ppm and 526 ppm. Rising smoke decreased the percentage of hydrogen replacement. The level of smoke was from 0.1 BSN to 4.9 BSN compared to BSN without EGR from 0 to 2 BSN. It shows that the average smoke emissions of EGR decreased while emissions of GHGs such as Co decreased by 20% EGR, by 0.04% to EGR by 0.74% and by 0.02% to EGR by 0.09%, while emissions of HC decreased by 12% to 19% EGR and EGR by 20% EGR [17].

Saravanan N et al. [34] In addition, Diethyl ether is achieved with carburizing and nozzle injection by four-stroke vertical diesel-fueled compression ignition engines and hydrogen. The optimal period for the injection of hydrogen was 5o BTDC, while thermal efficacité for both carburetion and hydrogen injection was respectively increased to 26.2 and 27.3 per cent. Higher thermal output was demonstrated by manifest hydrogen injection than the hydrogen injection combustion cycle.Smoke levels were lowered from 2 BSU to 0.8 BSU and further decreased to 0.7 BSU with the addition of hydrogen DEE. Carbon emissions tend to remain the same, not modified in volume by the 0.316 augmented modes of plain diesel and diesel-hydrogen. In a dual combustible DEE-hydrogen, however, the same CO dropped by 0.77 to 0.64 percnet by volume CO_2, by diesel hydrogen to 0.15 percent by volume CO_2, and by DEE–78 percent by hypoxygen packed. At 78 percent load, the maximum pressure was assessed. The peak pressure was 78.5 bar for pure diesel, but the average pressure remained the same, also in the double form of fuel (diesel + hydrogen) [17].

Gomes Antunes J M et al. [35], Have done an experiment with a single-cylinder, air-cooled diesel engine that naturally aims at managing the different hydrogen injections and even direct injectors in bi-fuel operations. The preheating of double fuel-air is not required since the pilot fuel is adequate to ignite hydrogen. In a direct hydrogen injection device, in order to ensure sure the hydrogen is ignited, the causing air will be heated up to 120^0C. For compression gasoline, the complete sovereignty of hydrogen is also hot. The engine capacity was produced by up to 42.8 percent pumped hydrogen compared with traditional diesel by 27.9 percent. The BTE was also reduced for EGR, though. The BTE decreased to such an extent relative to the EGR flow volume. The usage of BSEC without EGR is restricted, and the volume of EGR increases similarly. The BSEC is due to the double hydrogen process, which results, irrespective of the adverse effect of EGR on combustion, in higher calorie loss and improved engine performance. [17].

Szwaja Stanislaw et al. [36], Emphasized in the diesel/hydrogen dual-fuel configuration, their immense contribution and successes in the HCCI process of hydrogen combustion. This HCCI configuration has modified the mixture compression to 16. General hydrogen— the HCCI fuel combination helps the internal combustion engines to blast terribly. EGR and lean fuel mixtures also reduced the knock effect.HCCI and hydrogen aid in burning and starting combustion at various thermal settings of the motorhead and piston surfaces of 6^0-15^0CA bTDC, producing the highest tension before the TDC at around 50 percent mass burnt fraction (MFB) until the piston enters the TDC [37]. The experimental findings of the Kirloskar diesel engine were reported on 5.2 kW single-cylinder. The hydrogen of two type gas was measured with and without EGR. The injection was termed a Controlled Funnel Injection for the usage of 0.15 kg per hydrogen.The gross hydrogen mass flow rate was increased by 12.9 percent without EGR. The dilution of EGR at the same pace as hydrogen raising the BTE.

Saravanan N, et al. [38] After exhaustive experiments in Kirloskar AVL with 4-stroke vertical hydrogen compression petrol, a dual fuel system came to be. He examined dual fuel mode hydrogen integration with the air collector using combustion, operated port and multiple injection tools. The optimal injection hydrogen range is 5^0bTDC and 30^0CA, with numerous injections, respectively. The output of diesel brake heat is 23 percent, with timed injections (TPI), 26.2 percent, and engineered timed injections (TMI) at 27.3 percent.Compared with base diesel, NOx levels rose by 34% and 8% relative to TPI. But with overall CO pollution, the usage of SCR smoke has reduced dramatically to virtual nil. In the two carburizing hydrogen injection systems and the two TPI systems timed port injection was considered to be the most powerful. However, the NOx volume increased with TPI, but the ammonia flow to nitrous oxide (ANR) was decreased by 74 percent with specific catalytic reduction techniques (SCR).The average drop in HC with SCR was 0.9 to 73%. Whereas CO was down by 33% and carbon dioxide was up by 10%. CO_2 decreased due to carbon dioxide oxidation in the SCR cycle. The dual supply of hydrogen-diesel fuel provides a strong performance match and rising pollution with the exception of NOx. We find that the rise in NOx can be reduced in an SCR framework organisation [10].

Murari Mohan Roy et al. [39], The CI engine is fueled with traditional diesel fuel and replaced with petroleum hydrogen up to 90% of the overall energy consumption. The testing is performed by 4S, a mono cylinder, liquid-cooled, super loaded, constant rpm. Thermal efficiency increased by 0.3 by 42 percent with the fuel to air equivalence ratio. For the recirculation of 50% of the exhaust gas (EGR), the NOx was lowered by 60% without the remaining requirements. CO and HC emissions were impossible to find. Hydrogen may be applied to 90 percent of the overall power.Hydrogen was used by fast, uncompressible, and leaner fuel-to-air motors in an overloaded double motor. The hydrogen operation had a complete usable maximum pressure (MAP). The combustion between two phases was considered to be a higher degree between engine energy, and an indicator was established. The combustion was much quicker than expected in the full IMEP (with powerful two-stage combustion). EGR was designed to avoid the knock by increasing the temperature of the fuel [17].

Bari S. and Mohammed Esmail [40], Had studied the effect of small quantities of H2/O2 mixtures on the performance additive of 4 cylinder diesel engines. The required mixture quantification was accomplished by extracting hydrogen and oxygen from water electrolysis on board. The thermal blockages of 19 kW in this area were decreased by 2.6%, 2.9%, 22 kW, and 1.6%, in 28 kW. The average H_2/O_2in gasoline fuel was decreased by 6.1%.The engine dropped by 7.3% at 19 kW, 22 kW and 28 kW at 8.1% and a 4.8% average. NOx increases were reduced, and the HC, CO_2 and CO levels decreased. The strong average hydrogen fire rate near to TdC enhance combustion by up to 5% of the overall diesel energy in the form of H_2/O_2 absorption and subsequent increases in combustion.

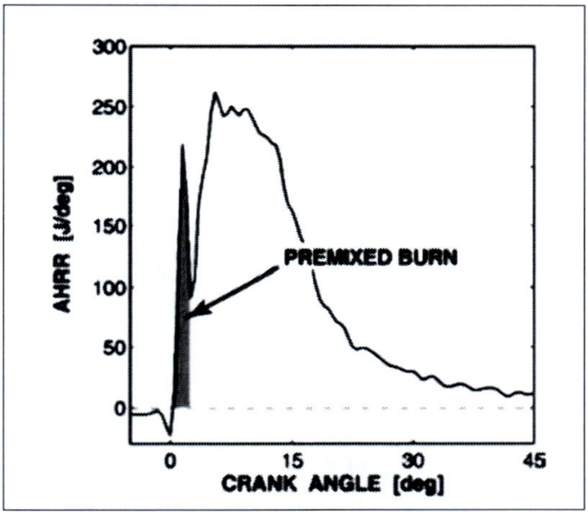

Figure 11. HRR vs. crank angle concentrated on pre-mixed combustion.

Jacob Wall [41] The thermal efficiency of the brake is projected to boost dual-fuel operations with diesel as the pilot fuel and 8 percent of additional power supplies with mass hydrogen. HC, CO and NOx hydrogen levels have been lowered to 8 percent appropriate. Gatts T et al. [42], Extensive testing carried out in 2004 found that the efficiency of the exhaust gas recycles turbochargers MP7, 365 E, 6 bikes, 265 kW and the variable configuration has improved to an ideal hydrogen level of 6%. The thermal braking impact at high loads has risen instead of the component load [17].

William F Northrop et al. [43] A detailed work was performed with multi-cylinder DI, 100 horsepower stop, General Motors developed a petroleum car and ULSD and soya (Glycine Max). Extensive check findings have been obtained. The simple measurements of hydrogen emissions for low-temperature combustion (LTC) equivalence ratio were tested with 0.5 to 1.1. Such contaminants are wide and should be managed via the techniques Lean NOx Traps (LNT) and SECR. Major differences have risen in toxins of Sulfur Diesel (ULSD) and B100. The inference is that the methods and processes of hydrogen pollution are controlled primarily by combustion control, but not by chosen fuels. The optimum equivalence ratio was found to be above 0.85 when the hydrogen emission was calculated [17].

Gregory K Lilik et al. [44] Direct injector, turbo-powered, light-duty diesel vehicle experiments are carried out with a large range of 103 kW DDC. Extra hydrogen has been taken into account on an energy base of 0% 2.5%, 7.5%, 10% and 15% for the vacuuming system's use. The extension of the injection period was necessary to increase NOx either through the enrichment of hydrogen or by the usage of baseline fossil fuels.For hydrogen injection, the equilibrium of NO and NO_2 has been altered to boost NO pollution and NO_2 pollution. These studies have confirmed and replicated the same findings using the PDF and CFD models distributed likelihood density and recognized that the NO_2 and H_2O hydrogen levels alone would not improve [10, 17].

Chong J J et al. [45] The Lister Petter TR1 studies, with 8.6 kW calculations, are carried out through a pump, a direct injection and a diesel motor. However, during a catalytic decrease in the α-aluminium emission process and the modified exhaust gas grid (REGR), 2% by weight is applied to ULSD in the form of CO and hydrogen. Complete, nitrogen monoxide (NO) accounts for 85-95% and nitrogen dioxide (NO_2) for 5-15%. The introduction to the engine intake range of hydrogen or type (H_2 CO and EGR) suggests overall amounts ofNOx are significantly reduced.The results have discovered several important parameters, including air/fuel ratio, piston combustion temperature and effect on nitrogen pollution, which affected therapeutic, preventive steps, such as special SCRs and diesel particulate (DPF) filters. Reformatting (H_2, CO and REGR) can lower rates of NOx and increasing the amount of NO_2 emissions. The REGR ratio of NO_2/NOx is higher as opposed to normal diesel operation. The usage of 15% hydrogen to remove NO_2 from NO is well known to be more effective with 10% of CO through REGR. The usage of DOOCs would also boost CO and HC pollution [10].

Liew C et al. [46], The heat release of the 1999 Model Cummins ISM 370, 10.8 litre size, 6 cylinders direct-injection, 276 kW diesel engine, two-stage combustion cycle was observed when operating on pure diesel, conducted extensive testing and demonstrated major achievements in the combustion area, in particular. The addition of hydrogen was observed to raise the RRH as seen in Figure 12 relative to pre-mixed combustion with the application of hydrogen to disperse fire.

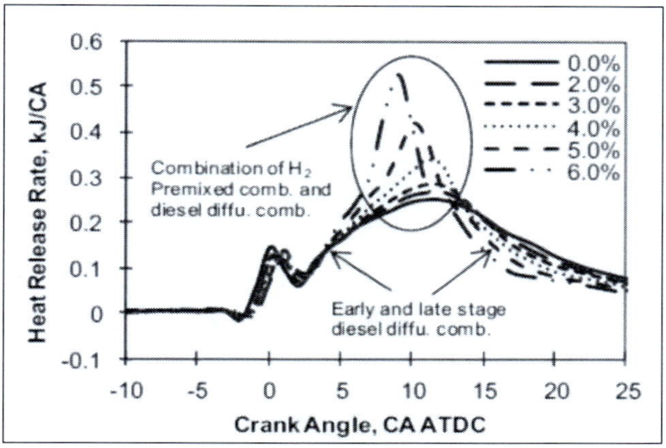

Figure 12. HRR vs CA.

Hohgsheng Guo, et al. [47], Experimental results indicate that the enriching of carbon-fuel fuel by hydrogen enhancement allows a HCCI diesel engine to work at higher and higher power levels. It increases the heat capacity of a braking device. B50 blends, peak tension and HRR are increased for ultra-low sulfur diesel (ULSD). The enrichment of hydrogen not only lowers total direct carbon dioxide and HC emissions but also decreases CO production per unit weight. However, enhanced hydrogen emissions from burnt machine diesel do not impact HC emissions significantly.NOx levels have not, though, modified significantly. The emission results showed that the combustion cycles with better hydrogen slows and reduces the combustion of HCCI diesel engines. Lower cetane number of fuels need additional hydrogen to prolong the combustion in optimum values. In fact, the combustion capacity of HCCI diesel engines at the time of use is improved by hydrogen-enriched. The possibility of knock will occur as hydrogen enrichment rises.

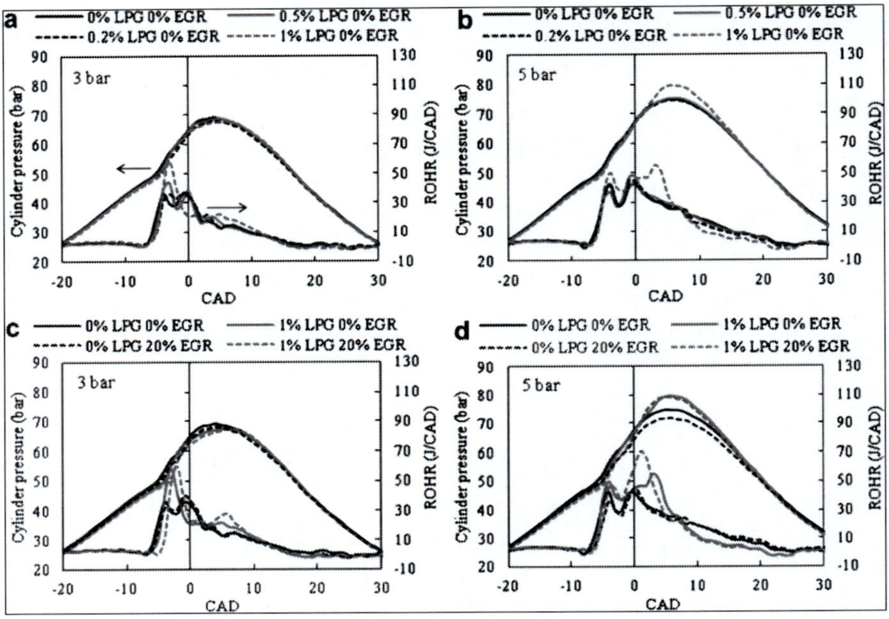

Figure 13. HRR vs. cylinder pressure at different conditions.

Tira H S et al. [48], LPG-Experimental fossil fuel fossil (LPG) –This research explores impacts on the combustion characters of a single-cylinder,4-stroke, 773 cc, 8.6 kW direct-injection engine of the direct injection diesel fuel such as rapeseed (Brassica napus), methyl ester(RME) and gas-to-liquid (GTL).

Ultra-low sulfur diesel (ULSD) cylinders and fuel escape rates (LPG) with specific LPG ratios. When GPL is removed, the ignition delay should be increased with a smaller volume of cetane and reduced heat costs. Through injecting LPG with the air into the piston, it allows the creation of a pre-mixed fuel mixture with low engine charges feasible in two combustion modes. The lag was reduced at the higher engine stage as opposed to the low speeds, and liquid fuel was pumped. The higher cylinders are used to promote strong engine load LPG combustion. In addition, as seen in Figure 13, greater heat intensity raises peak stress and the diffusion duration of combustion.

Zhou J H et al. [49], 88 kW of complete check, ISUZU 4EFI 4,3 Litre device, ultra-low sulfur diesel (ULSD), palm (Elaesis Guineensis) and 50% ULSD+ 50 percent PME Mixed gas at a constant velocity and with different loads of natural intake of hydrogen. The average hydrogen share was up to 20% of the overall fuel capacity. The effect on fuel used and the BTE depends on the load of the engine for all three chosen fuels with hydrogen supplementation.The BTE was partly charged in three fuels and strengthened at peak load for ULSD and ULSD + PME (B50), owing to the increased hydrogen presence in combustion. The shift was negligible for BTE with the SME. Yet hydrogen fuel substitution is 10% to 20% in palm methyl ether (PME).

Hohgsheng Guo et al. [50] HCCI supplements with a diesel capacity of 612 centimetres, variable n-heptane compression motor, have enhanced their function. The findings show that thermal efficiency has been increased by the streamlined combustion cycle, with the usage of a sufficient amount of hydrogen in pilot gasoline. Although hydrogen volume is adequate, the stated unbraked HC is decreased, the unbricked HC per unit of n-heptane mass is decreased gradually due to hydrogen absorption, and there would be more comparatively high tempo HC during the n-heptane kinetic cycle.Owing to the competition of some primary radicals, hydrogen is determined by the oxidation of the remaining HC during the high-temperature cycle. However, in internal combustion engines, the usage of hydrogen has its advantages; the same hydrogen, in the following section, has specific problems related to unfavourable combustion phenomena.

UNDESIRABLE COMBUSTION PHENOMENA

The characteristics of this hydrogen as the most reliable substitute fuel for IC Engines often lead to erratic combustion, mainly though it's not properly treated. Particularly since the inflammatory strength, high flame and maximums for undesirable phenomena are not typical; this can be summarized by unusual flammable anomalies such as [51].

Pre-Ignition

High flammability and low ignition strength of reactive hydrogen pre-ignition are sometimes found in the hydrogen driven internal combustion engine. The pre-ignition of the next process triggers higher combustion chamber temperatures because the combustion chamber has hot spots. This persists or extends until a pulse that gives energy back is eaten. The lowest ignition energy is calculated by the stoichiometric ratio as shown below.

Additional engine working environments in the combustion chamber is more susceptible to pre-ignition at high speed and load. Figure 15 displays result from an average hydrogen pressure level (IMEP) of 7 bar on a single pump, a 3200 rpm engine hydrogen check turbine. Pre-ignition details are shown with a dark line, and a dotted line was shown with the same engine with usual activity (regular combustion).In the case of pre-ignition, the peak pressure was higher than that of the standard combustion case; the same was progressed to 80^0CA bTDC. Therefore, there are no significant variations in intake pressures indicated by pre-ignition following the intake valves [52].

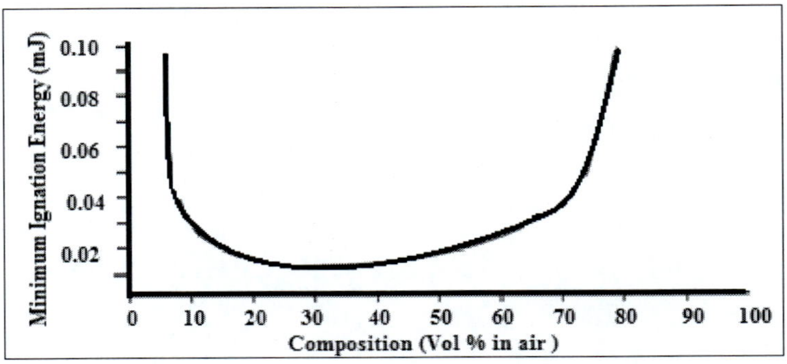

Figure 14. Minimum ignition energy of hydrogen in the air as a function of air composition by % volume.

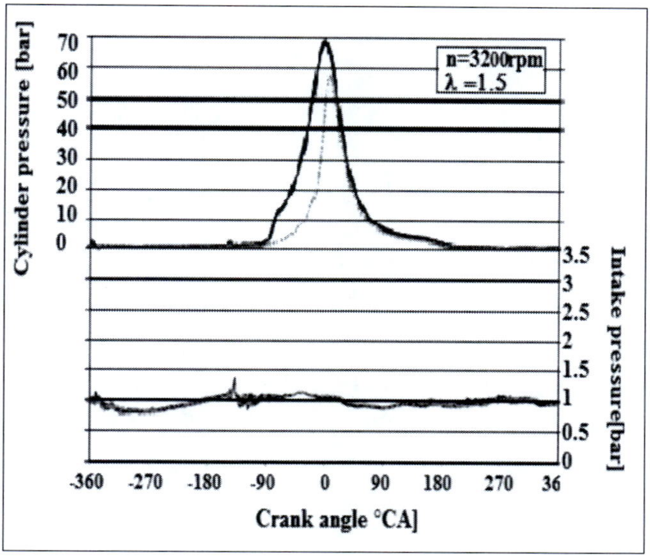

Figure 15. Typical cylinder and intake pressure [30].

Backfire

The explosion is a devastating product of a pre-ignition cycle. The ignited charge going through a portion of the valve and the inlet collector leads to the rear feather triggered by the inlet valve triggering when it is free. The back-fire is particularly evident in dual-fuel engines if the inlet collector triggers the hydrogen. The most significant distinction between back and pre-ignition is the length of the event. In locking place, valves are swallowed open, and rear fire experiences are usually pre-installed during the compression stroke.The frequency of the back-fire will not only inflict significant severe harm to the inlet structures of the engine. Pre-ignition is usually a precursor of backfiring. The safest approach is to configure the injector technique with the cumulative impact of a variable intake and outfeed valve timing [52].

Knocking

The temperature, time and pressure of the hydrogen-air blend trigger the spontaneously ignited charge, which contributes to the fast release of the heat which burns the available mixture, resulting in a wave called knock with a high amplitude pressure. This produced pressure wave will increase the engine's mechanical and thermal stress. Moreover, knock depends on the configuration of the engine and the mixing properties of fuel and air [52].

Hydrogen infusion or injection into the engine plays a crucial function in the production and regulation of these irregular combustion irregularities. Therefore, the rear portion explains different hydrogen activation or injection techniques.

Different Hydrogen Induction or Injection Techniques

The mechanism or device or technique for triggering or injecting water into the engine cylinder may be implemented to ensure an optimal existence of hydrogen in a combustion chamber and unintended anomalies of combustion, including pre-injection and retrofire. To order to establish or detect operating hydrogen abnormalities on the internal combustion engine, hydrogen induction or injection became a significant consideration.

There are 3 different techniques of hydrogen supplementation available from the literature [53, 54].

1. Continuous Carburetion (CC)
2. Manifold Induction or Injection (MPI) and
3. Direct cylinder Injection (DCI).

Continuous Carburetion (CC)

Carburetting is a process by the use of conventional gas carburettors which is most straightforward, oldest and known. Yes, this approach was favoured by the petrol engine due to minor improvements to the induction system. The advantage of this hydrogen induction is only drawn or decreased by using drawn or cross-drawn procedures and does not require high supply tension.

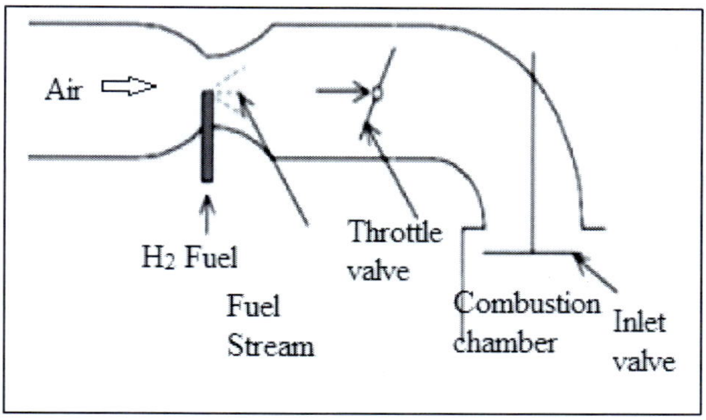

Figure 16. Continuous carburetion technique.

Earlier research on hydrogen induction was based on carburetion. Ricardo [55] and King [56-58] The hydrogen injection carbohydrate strategy has been implemented. Ricardo found that he "flipped off to a carbohydrocarbon," though he wanted a 3.8-compression engine as low as he could. While King always mostly concentrated on the same subject, he also had extreme unwanted combustion phenomena, such as pre-ignition and backfire. He also described the key causes for such unwelcome phenomena as free-flowing carbons and hot spots. Due to this issue, several more researchers later attempted to find specific inductions for alternative fuel. In such case, Swain & Adt [59, 60] supplied the hydrogen through a passage provided through the inlet valve.

Further, they regulated the hydrogen pressure by varying the hydrogen flow rate. Bindon J et al. [61], had searched for an outstanding strategy to provide standard managed to mix with a lean-burn carburettor. The carburization cycle has the negative side of pre-allusion and backfires because hydrogen/air are created by the inlet collector.

Manifold Induction or Injection (MPI)

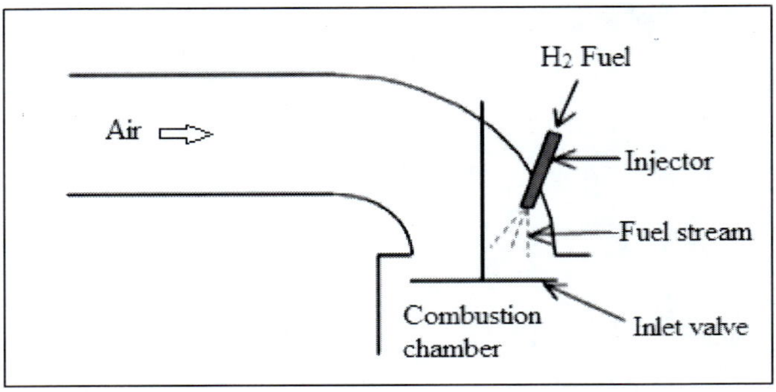

Figure 17. Manifold induction or Injection technique.

Gaseous hydrogen is provided either by mechanical or electrical injectors in the inlet numerous or ports injection techniques. The air is typically caused by the port or inlet collector during suction in multiple induction/injection [62]. Because in contrast to the central carburetion injection, the hydrogen volume is smaller, in this situation, the frequency of pre-ignition and backfire is reduced. Throughout this process, it remains necessary to adjust the quantity of air pumped and to control the energy generated by adjusting the hydrogen inducted/injected into an airstream. Two approaches include multiple injections, Synchronized Manual Injections (TMI) and continuous multiple injections are also possible.In the CMI method, hydrogen is pumped continuously throughout the intake time of the crank angle. Nevertheless, the backfire is more probable in this case, either because of the turn overlap or because of the speed or load shifts. With TMI, the duration of the injection and the velocity, that allows for hydrogen injection at the end of the suction stroke, can be controlled.

Lynch F E [63], Experienced with a special copper piping on simultaneous hydrogen infusion in the air intake system. He could pursue the backfire to a degree. Varde et al. [64] Successful inspection with an electronically operated injection of fuel near the intake valve is in the open condition of the valve was conducted for a specified time. Mac Carley et al. [65], Adopted and operated port injection with air entry but also taken specific engine parameters into account to avoid backfire.Karim et al. [66] used delayed port injection at low pressure to avoid the backfire. Majorly, Das L M [53] done exhaustive experimental investigations and studies on various hydrogen induction techniques and concluded that TMI is the best among other techniques available.

Direct Cylinder Injection (DCI)

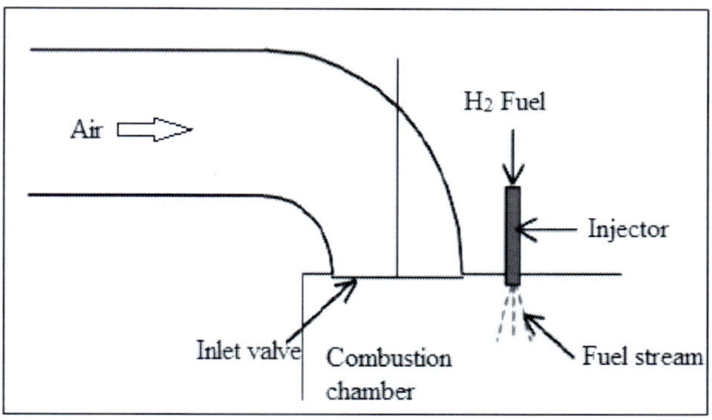

Figure 18. Direct cylinder injection technique.

More categorization was required in the LPDI (Low-Pressure Direct Injection) and High-Pressure Direct Injection Units. For all cases,hydrogen is immediately supplied to the combustion chamber. LPDI was supplied with hydrogen at the beginning of the stroke, and HPDI at the end of the compression stroke hydrogen was given [62]. The LPDI and HPDI are also contributing to the flushing of injector hot combustion materials and high-pressure materials when both device injectors are released in the combustion chamber, thereby bringing into doubt the injector's durability over time. Furthermore, indirect cylinder injection, the sufficient period for air exchange between hydrogen and air is a very limited source of weak hydrogen-fuel combination involvement in combustion [67]. The irregular combustion phenomena such as pre-ignition and rear fire can be prevented with direct gas injection techniques.

Erren [68], who used this concept of hydrogen injection directly into the combustion chamber in the year 1933 only. Homan H S [19] and [69], Late injection, rapid inflammation and (LIRIAM) technology, was developed to directly inject the hydrogen into the combustion chamber. Varde and Frame [64] and [70] Hydraulically powered injection of hydrogen was developed and used with direct cylinders. Timed multiple injections (TMIs) should be preferred between these processes for preventing irregular combustion, including pre-allusion and reverse fire, to increase further the required air intake hydrogen balance for better burning. Conventional diesel injectors are however not efficient and therefore not ideal for gaseous hydrogen transmission, due to the difference between gaseous and liquid fuel bulk modules.

Table 3. Different hydrogen injection techniques/methods [36][71]

Sl. No	Mixture formation	Classification	Hydrogen flow timing	Supply pressure
1	Continuous Carburetion	Pre-IVC	Continuous flow	A little above the atmosphere
2	Continuous Manifold induction/injection	Pre-IVC	Continuous flow	Slightly greater than the atmosphere
3	Timed manifold induction/Injection	Pre-IVC	The hydrogen flow begins after the intake valve is opened and ends before the valve closed	1.4 – 5.5 kg/cm^2
4	Low-pressure direct injection	Post IVC	The leakage of hydrogen starts after the valve intake closure and stops before the major pressure rise.	2-8 kgf//cm^2

Further details and comparison of these methods were shown in Table 3 [53].

REFERENCES

[1] Senthil Kumar M, Ramesh A, Nagalingam B. Use of hydrogen to enhance the performance of a vegetable oil fuelled compression ignition engine. *International Journal of Hydrogen Energy* 2003;28:1143-54.

[2] PS Ranjit NK, Mukesh Saxena et al. Studies on Combustion, Performance and Emission Characteristics of IDI CI Engine with Single-hole injector using SVO blends with diesel Asian Academic Research. *Journal of Multidisciplinary* (AARJM) May 2014;1:239-48.

[3] Ranjit PS, Sharma PK, Saxena M. Experimental Investigations on influence of Gaseous Hydrogen(GH2) Supplementation in In-Direct Injection (IDI) Compression Ignition engine fuelled with Pre-Heated Straight Vegetable Oil(PHSVO) *International Journal of Scientific & engineering Research* (IJSER) Oct. 2014;5.

[4] Ranjit PS. *Studies on Hydrogen Supplementation of SVO Operated IDI CI Engine for Performance Improvement and Reduction in Emissions.* Dehradun: University of Petroleum & Energy Studies; 2014.

[5] *National Hydrogen Energy Road Map- Path way for transition to Hydrogen Energy for India.* Ministry of New and Renewable Energy, Government of India.

[6] Japanese Ministry of Economy TaI. *Japan Hydrogen & Fuel Cell Demonstration Project.*

[7] Commission E. *The fuel cells and hydrogen joint technology initiative* (FCH JTI).

[8] Energy USDo. *Freedom CAR and Vehicle Technologies Multi-Year Program Plan* 2006–2011.

[9] S F. State of the art and future trends in hydrogen fuelled engines. *International Journal of Vehicle Design* 1983;4:359-85.

[10] PS Ranjit MS. Prospects of Hydrogen utilization in Compression Ignition Engines-A Review. *International Journal of Scientific Research* (IJSR) Feb 2013;2:137-40.

[11] Gillingham K. *Hydrogen Internal Combustion Engine Vehicles: A Prudent Intermediate Step or a Step in the Wrong Direction? Global Climate and Energy Project.* Precourt Institute for Energy Efficiency, Stanford University2007. p. 1-28.

[12] Das LM. Hydrogen-oxygen reaction mechanism and its implication to hydrogen engine combustion. *International Journal of Hydrogen Energy* 1996;21:703-15.

[13] White MC. SRR, Lutz EA. The hydrogen-fueled internal combustion engine: a technical review. *International Journal of Hydrogen Energy* 2006;31:1292-305.

[14] Overend E. *Hydrogen Combustion Engines.* The Unıversıty Of Edınburgh: School of Mechanical Engineering,1999. p. 1-77.

[15] Desert Co. *Hydrogen fuel cell engines and related technologies course manual, Module 3: Hydrogen use in internal combustion engines*.2009:1-29.

[16] Gupta BR. *Hydrogen fuel production, transport and storage:* CRC Press; 2008.

[17] P.S. Ranjit MS. A Review on hydrogen utilization in Internal Combustion Compression Ignition Engines. *International Journal of Science, Technology and Management* (IJSTM) Sept. 2012;3:1-21.

[18] Wong JKS. Compression ignition of hydrogen in a direct injection diesel engine modified to operate as a low-heat-rejection engine. *International Journal of Hydrogen Energy* 1990;15:507-14.

[19] Homan H S RRK, Deboer PCT, Mclean WJ. Hydrogen fueled diesel engine without timed ignition. *International Journal of Hydrogen Energy* 1979;4:315-25.

[20] Karim GA and Klat SR. *Experimental and analytical studies of hydrogen as a fuel in dual fuel engines.* ASME Paper No 75-DGP-9 1975.

[21] Karim GA KSR. Hydrogen as a fuel in compression ignition engines. *Journal of Mechanical Engineering* 1976;98:34-9.

[22] Gopal G RPS, Gopalkrishnan K V, Murthy BS. Use of hydrogen in dual-fuel engines. *International Journal of Hydrogen Energy* 1982;7:262-72.

[23] Naeser DB. The operation of dual fuel compression ignition engines utilizing diesel and methanol. *Proceedings of the IV International Symposium on Alcohol Fuels Technology*. Brazil1980.

[24] Haragopala Rao B, Shrivastava KN, Bhakta HN. Hydrogen for dual fuel engine operation. *International Journal of Hydrogen Energy* 1983;8:381-4.

[25] Ying-Qing LJ-DaL. *Improvement of combustion hydrogen fuelled engine*. New York: Pergamon Press; 1984.

[26] Weigang W. The research on internal combustion engine with the mixed fuel of diesel and hydrogen. *International Symposium Hydrogen Energy*. Beijing1985.

[27] Mathur HB, Das LM, Patro TN. Effects of charge diluents on the emission characteristics of a hydrogen fueled diesel engine. *International Journal of Hydrogen Energy* 1992;17:635-42.

[28] Naber JD, Siebers DL. Hydrogen combustion under diesel engine conditions. *International Journal of Hydrogen Energy* 1998;23:363-71.

[29] Saravanan N, Nagarajan G, Dhanasekaran C, Kalaiselvan KM. Experimental investigation of hydrogen port fuel injection in DI diesel engine. *International Journal of Hydrogen Energy* 2007;32:4071-80.

[30] Edwin Geo V, Nagarajan G, Nagalingam B. Studies on dual fuel operation of rubber seed oil and its bio-diesel with hydrogen as the inducted fuel. *International Journal of Hydrogen Energy* 2008;33:6357-67.

[31] Shirk MG, McGuire TP, Neal GL, Haworth DC. Investigation of a hydrogen-assisted combustion system for a light-duty diesel vehicle. *International Journal of Hydrogen Energy* 2008;33:7237-44.

[32] Saravanan N, Nagarajan G. An experimental investigation of hydrogen-enriched air induction in a diesel engine system. *International Journal of Hydrogen Energy* 2008;33:1769-75.

[33] Saravanan N NG. An Experimental investigation on performance and emissions study with port injection using diesel as an Ignition source for different EGR flow rates. *International Journal of Hydrogen Energy* 2008;33:4456-62.

[34] Saravanan N, Nagarajan G, Sanjay G, Dhanasekaran C, Kalaiselvan KM. Combustion analysis on a DI diesel engine with hydrogen in dual fuel mode. *Fuel* 2008;87:3591-9.

[35] Gomes Antunes JM, Mikalsen R, Roskilly AP. An experimental study of a direct injection compression ignition hydrogen engine. *International Journal of Hydrogen Energy* 2009;34:6516-22.

[36] Szwaja S, Grab-Rogalinski K. Hydrogen combustion in a compression ignition diesel engine. *International Journal of Hydrogen Energy* 2009;34:4413-21.

[37] Probir Kumar Bose DM. An experimental investigation on engine performance and emissions of a single cylinder diesel engine using hydrogen as inducted fuel and diesel as injected fuel with exhaust gas recirculation. *International Journal of Hydrogen Energy* 2009;34:4847-54.

[38] Saravanan N NG. An insight on hydrogen fuel injection techniques with SCR system for NOx reduction in a hydrogen diesel fuel engine. *International Journal of Hydrogen Energy* 2009;34:9019-32.

[39] Roy MM, Tomita E, Kawahara N, Harada Y, Sakane A. An experimental investigation on engine performance and emissions of a supercharged H2-diesel dual-fuel engine. *International Journal of Hydrogen Energy* 2010;35:844-53.

[40] Bari S, Mohammad Esmaeil M. Effect of H2/O2 addition in increasing the thermal efficiency of a diesel engine. *Fuel* 2010;89:378-83.

[41] Wall J. Effect of Hydrogen enriched Hydrocarbon combustion on emissions and performance. *Proceedings of the NPA2010.*

[42] Gatts T, Li H, Liew C, Liu S, Spencer T, Wayne S, et al. An experimental investigation of H2 emissions of a 2004 heavy-duty diesel engine supplemented with H2. *International Journal of Hydrogen Energy* 2010;35:11349-56.

[43] William F Northrop LMV. Investigation of Hydrogen emissions in partially premixed Diesel combustion. *Journal of Engineering for Gas Turbines and Power* 2010;132/112803-1.

[44] Lilik GK, Zhang H, Herreros JM, Haworth DC, Boehman AL. Hydrogen assisted diesel combustion. *International Journal of Hydrogen Energy* 2010;35:4382-98.

[45] Chong JJ, Tsolakis A, Gill SS, Theinnoi K, Golunski SE. Enhancing the NO2/NOx ratio in compression ignition engines by hydrogen and reformate combustion, for improved after treatment performance. *International Journal of Hydrogen Energy* 2010;35:8723-32.

[46] Liew C, Li H, Nuszkowski J, Liu S, Gatts T, Atkinson R, et al. An experimental investigation of the combustion process of a heavy-duty diesel engine enriched with H2. *International Journal of Hydrogen Energy* 2010;35:11357-65.

[47] Guo H, Hosseini V, Neill WS, Chippior WL, Dumitrescu CE. An experimental study on the effect of hydrogen enrichment on diesel fueled HCCI combustion. *International Journal of Hydrogen Energy* 2011;36:13820-30.

[48] Tira HS HJM, A. Tsolakis A, Wyszynski ML. Characteristics of LPG-diesel dual fuelled engine operated with rapeseed methyl ester and gas-to-liquid diesel fuels. *Energy* 2012;47:620-9.

[49] Zhou JH, Cheung CS, Leung CW. Combustion, performance and emissions of ULSD, PME and B50 fueled multi-cylinder diesel engine with naturally aspirated hydrogen. *International Journal of Hydrogen Energy 2013*;38:14837-48.

[50] Guo H, Neill WS. The effect of hydrogen addition on combustion and emission characteristics of an n-heptane fuelled HCCI engine. *International Journal of Hydrogen Energy* 2013;38:11429-37.

[51] Verhelst S. WT. Hydrogen-fueled internal combustion engines. *Progress in Energy and Combustion Science* 2009;35:490-527.

[52] Verhelst S. WT. *Hydrogen-fueled internal combustion engines.* Belgium: Ghent University, Sint-Pietersnieuwstraat; 2009. p. 1-80.

[53] Das LM. Fuel induction techniques for a hydrogen operated engine. *International Journal of Hydrogen Energy* 1990;15:833-42.

[54] Murat Ciniviz HK. Hydrogen use in internal combustion engine: a review. *International Journal of Automotive Engineering and Technologies* 2012;1:1-15.

[55] Ricardo HR. Further note on fuel research. *Proceedings of Inst Automobile Engineers* 1923;18:327-41.

[56] King RO HSV, Allan AB, Anderson RWP, Walker J. The hydrogen engine: combustion knock and the related flame velocity. *Transactions of Engineering Institute of Canada* (EIC) 1958;2:143.

[57] Rand ROKaM. The oxidation, decomposition, ignition and detonation of fuel vapours and gases-XXVII. The hydrogen engine. *Canadian Journal of Research* 1955;33:445.

[58] King RO WAWWA, Mahapatra B. The oxidation, ignition and detonation of fuel vapours and gases-V. The hydrogen engine and the nuclear theory of ignition. *Canadian Journal of Research* 1948;F26:264.

[59] Adt MRSaRR. The hydrogen-air fueled automobile. *Proceedings of 7th Intersociety Energy Conversion Engineering Conference*: American Chemical Society; 1972. p. 1382.

[60] Swain M R JMP, Robert R Adt Jr, William J D Escher, *Hydrogen-fuelled Automotive Engine Experimental Testing to provide an initial Design Data Base*. SAE 1981.

[61] Bindon J JH, J. Simmons, P. Mahlknecht and C. Williams. The development of a lean-burning carburettor for a hydrogen-powered vehicle. *International Journal of Hydrogen Energy* 1985;10:297.

[62] Saravanan N, Nagarajan G. Performance and emission studies on port injection of hydrogen with varied flow rates with Diesel as an ignition source. *Applied Energy* 2010;87:2218-29.

[63] Lynch FE. Parallel induction: A simple fuel control method for hydrogen engines. *International Journal of Hydrogen Energy* 1983;8:721.

[64] Varde KS FGA. A study of combustion and engine performance using electronic hydrogen fuel injection. *World Hydrogen Energy Conference* - IV1982.

[65] Vorst CAMaWDV. Electronic fuel injection techniques for hydrogen-powered engines. *International Journal of Hydrogen Energy* 1980;5:179.

[66] Karim GA and Watson HC. Experimental and computational considerations of the compression ignition of homogeneous fuel deoxidant mixtures. *SAE transactions* 1971;80:450.

[67] Masood M. IMM, Reddy SA. Computational combustion and emission analysis of hydrogen–diesel blends with experimental verification. *International Journal of Hydrogen Energy* 2007;32: 2539-47.

[68] Erren HCWaRA. Hydrogen: A commercial fuel for internal combustion engines and other purposes. *Journal of institute of Fuel* 1933;VI:277-91.

[69] Homan HS. *An Experimental Study of Reciprocating Internal Combustion Engines Operated on Hydrogen.* USA: Cornell University; 1978.

[70] Varde KS, Frame GA. Hydrogen aspiration in a direct injection type diesel engine-its effects on smoke and other engine performance parameters. *International Journal of Hydrogen Energy* 1983;8:549-55.

[71] Das L, Gulati R, Gupta P. Performance evaluation of a hydrogen-fuelled spark ignition engine using electronically controlled solenoid-actuated injection system. *International Journal of Hydrogen Energy* 2000;25:569-79.

In: Applied Soft Computing Techniques … ISBN: 978-1-53618-180-7
Editors: Amit Kumar Thakur et al.

Chapter 4

SAFETY ASPECTS IN HANDLING AND STORAGE OF ULTRA HIGH PURITY GASEOUS HYDROGEN

P. S. Ranjit*[*], *PhD
Department of Mechanical Engineering,
Aditya Engineering College (Autonomous),
Surampalem, Andhra Pradesh, India

ABSTRACT

Hydrogen is typically preserved in two ways and provided as compressed hydrogen or cryogens. The most growing way of storing hydrogen is steel or composite cylinders/tanks in different sizes and capacities. Often they may be packed in a box or deposited for transport in a basket known as a cascade. Owing to its tiny molecules, hydrogen is prone to spill easily into some brittle structures, fractures and/or poor joints in storage tanks rather than certain traditional gasses under similar strain. Though hydrogen is typically non-corrosive and does not interfere with the storage materials in vessels, it may absorb material

[*] Corresponding Author's Email: psranjit1234@gmail.com.

fragmentation at the grind structure. Additionally, a loss of hydrogen protection can happen in the event of fires and the construction content used in storage tanks will be damaged. In the worst case scenario, a hydrogen storage tank that generates a blow wave and flight projectils/missiles could result in a catastrophic failure. Therefore, good quality requirements must be established and adhered to in the hydrogen storage facilities to ensure container integrity. This chapter provides an analysis of the main safety and technological issues relevant to the decisions created for hydrogen storage. This should be pointed out that, as this technology is more widely used and frequently utilized–unenergized discharges, fires, and their containment capacity and handling for domestic combustion engines–this chapter primarily concentrates on hydrogen storage systems.

Keywords: gaseous hydrogen, safety, NFPA Class I Division 2 Group B standards, use in I C engines

INTRODUCTION

The smallest and lightest hydrogen molecule accumulates if the gas is released on the roof of the encircled zone. Hydrogen diffusiveness is ten times greater than gas and other gasoline vapours [1-3, 4].

The equivalence ratio of = 1.8 reflects the burning rate of the hydrogen-air mixture as = 1.1 results in a hydrocarbon air mixture [5]. The researchers must consider all the safety issues needed to deal with high-pressure gas hydrogen. The following is an overview of different issues:

Storage area safety aspects
Handling area safety aspects
Engine specific systems
Health-related safety issues

Maximum Allowable Quantity (MAQ): The maximum permissible hydrogen requirements in storage and use places.

The collection areas are referred to as gas boxes and empty container sprinkler and sprinkler.

Table 1. Hydrogen and conventional diesel fuel leakage profile

Sl.No	Description	GH_2	LH_2	Conventional Diesel Fuel
1	Representation of images			
2	Dispersing into atmosphere	Scatters easily as gas as	In the leaked location, puddles appeared.	Appeared puddles. Very little vaporisation.
3	Visibility	No	In fog and freeze, visible	In liquid form visible
4	Odor	None	None	Yes
5	Flammability	The Fire / Ignition really fast.	The Fire / Ignition really fast.	Comparatively slow
6	Flame	At daylight, invisible.	At daylight, invisible.	Visibility in smoke form

Table 2. MAQ specifications of the storage area

Sl. No	MAQ of GH_2 storage	Specifications		
		Sprinkled area	Gas Cabinet	Exhaust enclosure
1	< 28 m^3	No	No	No
2	>28< 56 m^3	No	Yes	Yes
3	>56<112 m^3	Yes	Yes	Yes

While the storage of gas is up to 28m^3 per amount, no need exhaust containers under a sprinkled area are essential for gas cabinets.

When the warehouse is filled with a sprinkler, no gas case and exhaust enclosures are needed for the storage of up to 56m^3 of Hydrogen MAQ [6].

Storage and Usage Area Construction

The content of' non-fuel' must be used to design and handle storage areas as specified by ASTM E 136 [7, 8]. The trail to these sites shall be from 20 feet to 50 feet long and have a height difference of 13.6 feet with a minimum road width for convenient fire protection equipment access to these sites in emergencies [9].

Exhaust ventilation for such buildings must be given up to 1 meter above the ceiling since hydrogen gas is lighter than the soil, absorbs the roof with hydrogen and adds to the occurrence with pollution. For mechanical ventilation the rate is not under 3048m^3/min/m^2 [10].

In all places in which compressed gases are processed, generated, or used, it is essential to use hazard recognition signs. And all of these indicators must be allowed to keep anybody within 25ft of the storage space in an access mode [11, 12].

Some of the signs are appended below [13].

WARNING: HYDROGEN – FLAMMABLE GAS
NO SMOKING – NO OPEN FLAMES

Table 3. Various signs of being available at the storage site

Sign	Indication	Sign	Indication
No Smoking 	The Identification of areas where smoking is prohibited	Gas Shutoff Valve 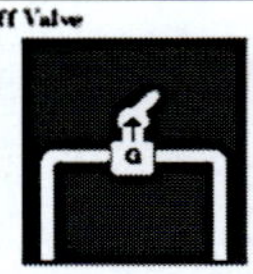	The location of gas shutoff valve
No Open Flame — Flame 	The identification of areas where the open fires are not permitted	Electric Panel or Electric Shutoff 	The identification and location of an electrical panel or other electric shutoff devices
Fire Extinguisher 	For everyday use in working places and public areas:		Gas detector

When a leak occurs, the hooter must alert of leaching, using the low explosion limit (LEL) and the hooter and warning system as shown below, by the shutdown of gas and electrical power as soon as possible [14-18]. One of these implemented gas sensors was shown below with its monitor.

Fire Extinguishers [1, 3]

Fire: Fire is a chemical reaction wherein the process of burning material is oxidized. The reaction is a chain reaction, an exothermic reaction.

For gases, the storage of fire extinguishers class B is essential, but class A is to be used to sustain solid fire such as coal, paper clothes etc. [13].

Electrification [1, 3]

Similar acceptable requirements are required according to the necessity of the use of high-pressure gaseous Hydrogen. Less than the MAQ, the application of the article is restricted to minimal hydrogen storage, and the storage capacity is 47 Litre of water at a pressure not approaching 140 Bar.

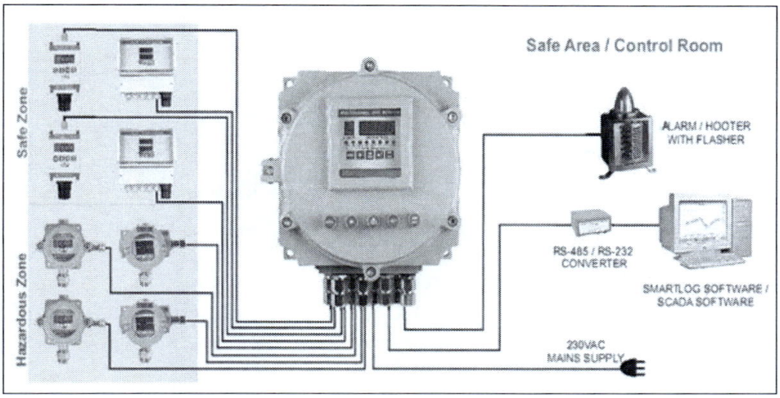

Figure 1. Clear visual view of the monitoring system for hydrogen gas leakage.

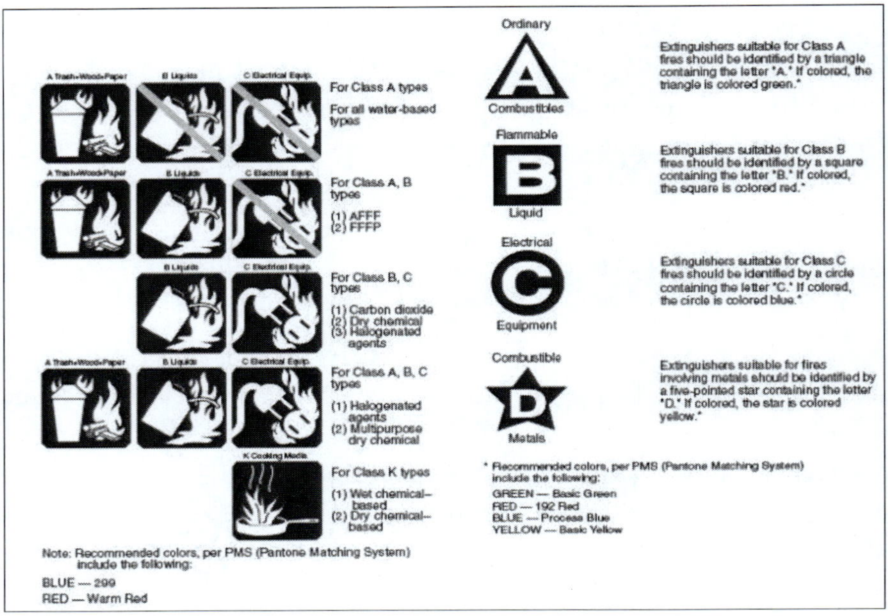

Figure 2. Fire extinguishers.

Therefore, NFPA regulation Group B section I, Division 2 is valid in this sector for procurement of electrification [19].

Class I shows the flammable air gasses are likely to inflammable or explosive in sufficient amounts.

Division 2 shows that in contained cylinders, flammable gasses are stored

and

Group B, Specify in sufficient quantities in the air the existence of inflammable gasses with a healthy laboratory distance below or equivalent to 0.45mm. Hydrogen is found in this group.

Wiring

ZHFR/HFFR wire [20] is used for rigid or steel intermediate ducts.

Luminaries

Suitable metal shutdowns are required as a potential inflammatory source at graded locations for lighting and other heat-generating devices, switches, circuit-breaker devices and connectors, as specified in Class I of Division 2 Group B.

Therefore, the luminaires must be protected against physical damage by adequate security or location. Where there is a possibility of small amounts of flammable vapours or gases being sparked by dropping sparks or hot metals of lamps and light bulbs, adequate bins are installed, and other useful precautions [21].

Protection Concepts [1, 3]

In these environments, different types of devices can be used to ensure the fire risk is reduced or eliminated substantially. The machinery must be designed and produced according to unique building specifications referred to as safety principles.

Flameproof – Ex d

Equipment that might cause an explosion in a region that can be explosive and prevent transport to an insecure environment. The safety strategy often prohibits toxic air from coming into contact with machinery and reaching the container.

Ingress Protection

A concern is a defence from penetration of sharp foreign objects and water while securing machinery in dangerous areas. This is classified as entry security and is commonly known as the I.P. address. IEC 60529 is the accepted norm for entry-level defense. And I.P. 65 is the perfect one.

Some of these electrical devices are appended below.

Handling must be segregated by a lot of walls, public roads, open flames, and so on.

Table 4. Details of protection concepts

Protecting Aspects	Equipment Code	Description	Standards (International Level)	Zone suitability
The goal is to avoid an inflammation out of the equipment	Ex d	Protection against flames	IEC 60079-1	1, 2

Table 5. Ingress progress 65 details

Representation of 1st Numerical	Degree of Protection	Representation of 2nd Numerical	Degree of Protection
6	Full security against dust intrusion	5	Protected from any distance from water jets

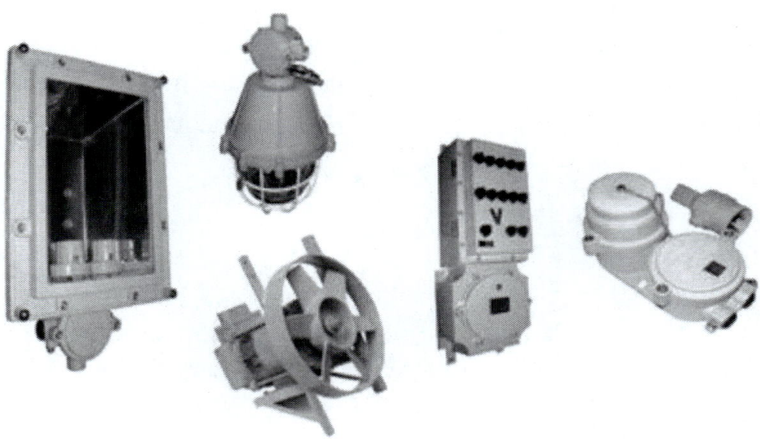

Figure 3. NEC of NFPA based electrical equipment.

Table 6. Maintained explosive distance

Sl.No	Description	Minimum distance maintained from storage space in feet
1	Lot Lines	45
2	Exposed persons other than those involved in servicing of the system	25
3	Air Intakes – Compressors	45
4	Unclassified Electrical Equipment	15
5	Utilities (overhead) including electrical power building services, Hazardous materials, piping	20
6	Ignition sources such as open flames and welding	45
7	Parked cars	25
8	Flammable Gas storage systems including other Hydrogen systems above the ground	20
9	Ordinary combustible including fast burning solids such as Ordinary lumber, excelsior, paper and vegetation other than found in maintained land scraped areas	20
10	Heavy Timber, Coal or other than slow burning combustible solids	20
11	Smoking	25

Table 7. Separation of cylinders by hazard class gases

Sl. No	Gas Category	Minimum distance maintained from storage space in feet
1	Toxic or High Toxic	20
2	Pyrophoric	20
3	Oxidizing	20
4	Corrosive	20
5	Unstable Reactive class 2, 3 and 4	20
6	Other Gases	No separation required

HYDROGEN HANDLING SYSTEM ISSUES [1, 2]

The Hydrogen contained as it is in containing cylinders must be planned, produced, inspected and labelled in compliance with standard test protocols, such as the ASME boiler [22-27]. You must ensure that these cylinders under no circumstances are heated to over 52^0C., and this cylinder must be protected to avoid physical injury. The cylinders should be equipped with standard safety precautions such as guard posts, chain belts etc. Whether the cylinders are kept empty or full, they should be used with Cylinders. And these hollow so full tubes have to be turned off. Never let the nozzle fail. If the valve is incapable, the whole pressure vessel can be converted into a projectile. One such event is described below [28].

Even for small distances cylinders are not to be pulled. For limited transport distances, manual gas trucks must also be utilized [29]. Never encourage mechanical shocks to occur with the cylinders. Cylinders should be found in the pamphlet Part-1 in compliance with either ISO Standard 11625 or CGA.

The pressure regulator is one of the key components needed for transporting the gas at a test stand of the engine from high to a low pressure on the workplace. Use model content of the SS 316 for the use of specially designed and produced two-stage regulators. The double-stage and single-stage pressure generator focused on gas demand. As shown below, during safety inspections by the regulator, the user must take into account safety aspects [30].

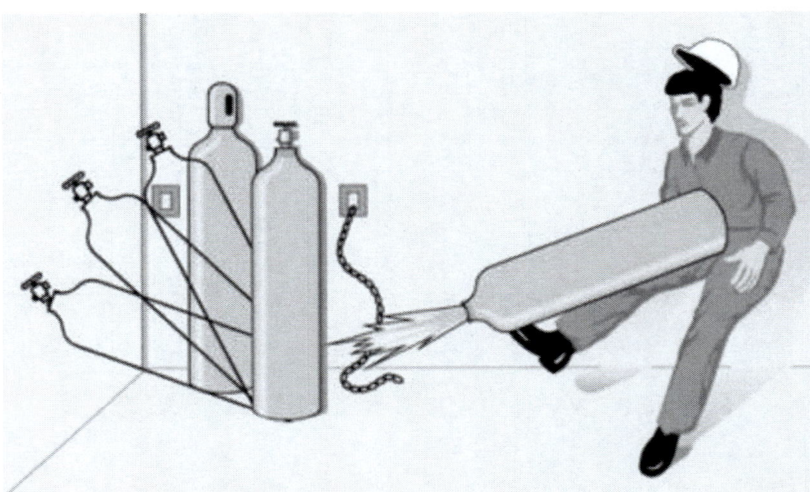

Figure 4. How does the compressed gas cylinder become a rocket [1].

Figure 5. Cylinders carrying trolleys.

Figure 6. Safe way to use pressure regulator.

The second most critical is the relief valve. When gaseous Hydrogen is not supplied at low pressure by the pressure regulator, the flow components of the Hydrogen in the flow line can be disrupted. If the gasses enter these damaged parts, the risk becomes immense. This relief valve is therefore necessary, and its outlet must maintain the air intact by damaging the controller that leaks high-pressure gas. The release mechanism and pressure relief valves must be operating according to the compressed gas association (CGA) G5, S 1.1 and NFPA standards [23, 24, 31].

The development and installation of tubing, valves and fittings following the standard NFPA and IFGC codes [25, 32, 33] are of great importance. Link just tube fittings always with the tube and several system/pieces/appliances, while stopping BSP and NTP threads when using gas hydrogen, as much as possible. It is also essential for qualified engineers to apply NFPA standards periodically, in accordance with the standard maintenance strategy. All records must be held for 3 years. Never use hydrogen storage and transportation components of Cast Iron material. SS 316 content for this gas is always preferred. All of these components and tube assemblies must be tested with a pressure test no less than 1 1/2 times the workforce, in accordance with the required standard procedures in the NFPA, IFGC and ASME standards [34].

ENGINE SPECIFIC SYSTEMS [1, 2]

In addition, the majority of scientists utilize two methods when integrating this Hydrogen in the reactor. These are injectors and direct implants, i.e., methods of activation and delivery. The main focus is on several injections. This multiple injections is closer and higher than the atmospheric pressure at the hydrogen delivery pressure in the inlet manifold.

This gas is generated due to the pressure difference between the internal and external environment. The reverse fire due to pregnancy and the heat location of the combustion chamber is a significant problem in this induction method.

In Delhi, the Indian Institute of Technology (IIT), much research has been carried out. So many induction methods, including carburetion, Timed Manifold Injections (TMI), are usable. Detailed study shows that Timed Manifold Injection is the safest induction method for handling unwanted combustion processes.

In the exhaust system, well-deployed water sprays would ideally be used to avoid the detonation by this cooled exhaust. CO_2 can also be frequently transferred to this exhaust system to control this unpleasant phenomenon.

Inert gasses, such as oxygen, helium and carbon dioxide extinguishers, such as ammonium phosphate and potassium chloride are ideal for use in test cells to mix Hydrogen with the air in a combustible area if there are hydrogen losses [33-37].

Also delayed ports are sometimes equally helpful to inject Hydrogen into the engine as is Timed Manifold Injection (TMI), a very safe process [38].

HEALTH-RELATED HAZARDS IN MAKE USE OF HYDROGEN [1, 3]

The potential health risks of this Hydrogen were high when the patient inhaled that the lightest gas is mixed with air in the confined area and replaced with oxygen, which includes asphyxiation while escaping. The victim can't also suffer this situation if he is asphyxiated. Because of this oxygen-deprived environment, the person can be aware of dizziness, diarrhoea, fatigue and mobility loss.

And there are no negative skin results. When this Hydrogen is being exposed to eyes, the eyes start irritating, then the patient must be quickly clean his or her eyes. If Hydrogen lasts for 15 minutes, leaked Hydrogen eliminates the source of oxygen. It is also very challenging to stifle it and to foresee the condition. The victim loses consciousness without awareness. The oxygen-deficient condition of the patient such as dizziness, salivation, diarrhoea and nausea may cause symptoms. However, there are no adverse effects exposed to the skin [39].

Marking of Hazardous Area Equipment as per IEC

All equipment for use in hazardous areas should be marked as prescribed in 60079-0.

Marking of Hazardous Location Equipment as per North American Standards

The marking of hazardous location equipment varies in accordance with the relevant Electrical Code. As a general rule equipment should be marked as follows:
- Company/Manufacturers name
- Class I, II and/or III
- Division 1 and/or 2
- Group A, B, C, D, E, F and/or G
- Approving NRTL (Nationally Recognized Testing Laboratory) logo
- Product identification
- Serial number
- Other relevant safety information
- Enclosure Type Rating

(A) Fire Extinguishers

S.No.	Description	Qty
1.	**Ceasefire Brand** ABC MAP: IS 13849, ISO 9001 & **CE certified** extinguishers and carry the certificates, the extinguishers manufactured as per IS s15683: 2006. Performance as per IS 15683: 2006, have **Fire Rating of 21A, 183 B**, extinguisher body made up of Mild steel and thickness of the body 2.0 mm, contain **Mono Ammonium Phosphate 90%** (100% pure)to net weight) as Fire fighting agent. Mono Ammonium phosphate powder discharge less than 14 sec. Activation time of the Fire Extinguisher 2 seconds, effective throw of the Extinguisher 13.5 ft., **5 year shelf life and warranted for complete life.** **5 Kg Capacity**	03
2.	**Ceasefire Brand** CO_2 extinguishers equipped with used – unused indicator ISI, ISO & DOE certified **4.5 Kg Capacity**	02
3.	**Ceasefire Brand** CO_2 extinguishers equipped with used – unused indicator ISI, ISO & DOE certified **22.5 Kg Capacity**	01

Hydrogen Gas Leak Detectors

Consists of Sensor Element housed in flame proof &
weatherproof housing (IP55) SS316, 2 Wire 4-20 mA
Intrinsically safe Transmitter PCB housed in FLP.
Junction box (Applicable for class II C)
FLP Cable glands - (Applicable for class II C)
Sensor Type: Catalytic
Calibrated for Hydrogen
Sensor Life: App. App. 5 – 8 years depending upon exposure of gas.
Detection Range: 0 – 100% LEL
Output – 4 – 20 mA,
Operating Pressure – Ambient
Response time at 20 Deg. C : T90< 10 sec
Storage Life : Unlimited
Supply of Transmitter : 24 VDC @ 300mA
Weight : Max. 500 gm

Hydrogen Gas Flame Detector

Explosion proof UV/IR Flame Detector
Make: Electron Standard Pribor, Russia
Model : IPES – UV/IR
Sensor Type: Ultraviolet/Infrared
Distance/Cone of vision: 25 meters/90 Degree.
Power Supply: 12-32V
Output: 4 – 20 mA, RS-485
Operating Temp. : -60 to +90 Deg. C.
Class of protection: IP66, IP67
Approval: CE/ ATEX/FM / CMRI / CCOE
Marking: Eex d IIC T4 II 2 G
Certificates: FM/EN54
Body Type: Aluminum
Wiring: 14 AWG(2.08mm) or 16AWG (1.31mm)
Cable Entry – ¾” -14 NPT
Weight – Aluminum 5.5 lbs (2.5kg)
Warranty – 5 Years

Further, Products needs to be approved from

Sl.No	Approving Authority	Description
1	CMRI, Dhanbad	The Remote gas Sensor Housing for continuous gas monitoring system has been certified for Explosion proof as per IS 2148 – 2004 (IEC 60079-1/2001) and Weather Proof (IP 55) as per IS: 12063-1987 [IEC 60529-1989] & 4-20 mA transmitter for Combustible Gas Detector is certified for Intrinsically safe as per IS5780:2002 / IEC 60079-11(1999) by CMRI Dhanbad.
2	BIS (Bureau of Indian Standards) Approval	The Remote gas sensor has been granted BIS certification marks Licence no. CM/L-7850082 as per IS 2148: 2004 by Bureau of Indian Standards, Mumbai
3	CCOE (Chief Controller Of Explosive) Approval	All the remote gas sensor for continuous gas monitoring systems are approved by CCOE, Nagpur vide Identification No.P226024/1 and Test Report No.CIMFR/TC/P/H852 Dt.29/01/2008

References

[1] Ranjit, P. S., MS. State-of-the-art of Storage and Handling issues related to High Pressure Gaseous Hydrogen to make use in Internal Combustion engines. *International Journal of Scientific and Engineering Research (IJSER),* Sept. 2012; 3:1 - 17.

[2] Ranjit, P. S., MS. Prospects of Hydrogen utilization in Compression Ignition Engines- A Review. *International Journal of Scientific Research (IJSR),* Feb. 2013; 2:137 - 40.

[3] Ranjit, P. S. *Studies on Hydrogen Supplementation of SVO Operated IDI CI Engine for Performance Improvement and Reduction in Emissions*. Dehradun: University of Petroleum and Energy Studies, 2014.

[4] Raj, P. K. *Clean Air Program: Design Guidelines for Bus Transit Systems using Hydrogen as an Alternative Fuel.* Technology and Management Systems Inc. for U.S. Department of Transportation, Federal Transit Agency; October, 1998.

[5] V. M. Hydrogen Safety Research: State-of-the-Art. *Proceedings of the 5th International Seminar on Fire and Explosion Hazards 23-27th*, April, 2007.

[6] NFPA. *Maximum Allowable Quantity (MAQ) of Hydrogen per control Area,* Table. 6.4.1.1. p. P.No. 2.26.

[7] ASTM. *Standard Test method for Behavior of Materials in Vertical Tube Furnace at 750 deg*. C. ASTM E 136 Standards.

[8] Code NBCaS. *Construction Types and Height and Area Requirements*, Chapter 7, Section: 7.1.4.1.

[9] Code NBCaS. *Bridges*, Chapter 7, Section 7.1.5.2.5.5. NFPA. p. 91.

[10] Code NCGaCF. *Exhaust Ventilation*, Chapter 6, Section 6.4.2. p. 17.

[11] Code NCGaCF. *Hazard Identification Signs*, Section 6.11.1, 6.11.2.1. and 6.11.2.2.

[12] NFPA 704: Standard System for the Identification of the Hazards of Materials for Emergency Response. *Flammability Hazards*, Chapter 6, P. No. 704 - 8, Special Hazards.

[13] 170 N. Symbols for General Use, Chapter 4, *Standard for Fire Safety and Emergency Symbols*. p. 13,4,5,7,30, 3 and 50.

[14] 70 NEC. *Combustible Gas Detection System*.

[15] Explosive Atmosphere.

[16] *Performance Requirements of Detectors for Flammable Gases*. In: 2075 AU, editor.

[17] ANSI/ISA 60079-29-2- *Explosive Atmosphere P. Gas Detectors-Selection, Installation, Use and Maintenance of Detectors for Flammable Gases and Oxygen*.

[18] 12.13.03 IT. *Guide for combustible Gas detection as a method of protection*.

[19] NFPA 70 National Electric Code AHCLCI, II and III Division 2. *Special Occupancies, Electrical Wiring and Equipment*. p. 70 - 369.

[20] NFPA 70 NEC, Article 501. *Wiring, Class I, Locations*. p. 70 - 376.

[21] NFPA 70: *National Electric Code A*, Section: 501.1/5 (B) (1) through (B) (4). Switches.

[22] Gases NCC. *Containers, Cylinders and Tanks*, Section. 7.1.5.1.8. p. 55.

[23] CGA S.1.1.- Part 1 CfCG. Pressure Relief Device Standards-Compressed Gas Association.
[24] Code NCGaCF. *Pressure Relief Valves,* Chapter.7, Section. 7.1.5.5.2. p. 21.
[25] Code NCGaCF. *Design of Gaseous Hydrogen Systems*: Section 10.3.2, Chapter 10 Gaseous Hydrogen Systems. p. 40.
[26] Lee, C. Cadwallader, K. G. D., J. Stephen Herring. *Hydrogen and Oxygen Gas Monitoring System Design Operation-Safety and Technology of Nuclear Hydrogen Production, Control and Management.* Idaho National Laboratory, June, 2007.
[27] Air Products and Chemicals Inc. *S-. Gaseous Hydrogen*. Air Products and Chemicals Inc., 7201 Hamilton Boulevard, Allentown, PA 18195 - 1501; 2010. p. 5.
[28] Air Products and Chemicals I, Safetygram-14. *Don't Turn a Cylinder Into a Rocket*. 1998. p. 2.
[29] Air Products and Chemicals Inc. S-. *Handling, Storage, and Use of Compressed Gas Cylinders.* 2004. p. 4.
[30] Air Products and Chemicals I, Safetygram-12. *Regulator Selection, Installation, and Operation*. 1998. p. 4.
[31] Code NCGaCF. *Hydrogen Vent System,* Section.10.2.2, Chapter 10. p. 39.
[32] Air Products and Chemicals I, Safetygram-23. *Cylinder Valves.* 2007. p. 8.
[33] Code NCGaCF. *Physical Protection,* Chapter 7, Section 7.1.8.3.1, 7.1.8.3.2., 7.1.8.3.4, p. 22.
[34] Code NCGaCF. *Testing of Hydrogen Piping systems*, Section 705.1., 705.3.1., 705.3.2 and 705.4.1. p. 34.
[35] Das, L. M. Safety Aspects of a Hydrogen fuelled Engine System Development. *International Journal of Hydrogen Energy,* 1991; 16: 619 - 24.
[36] Das, L. M. On -board Hydrogen storage systems for Automtoive Application. *International Journal of Hydrogen Energy,* 1996; 21: 789 - 800.

[37] Das, L. M. Hydrogen Engines: A View of the Past and Look into the Future. *International Journal of Hydrogen Energy,* 1990; 15:425 - 43.

[38] Das, L. M. Fuel Induction Techniques for a Hydrogen operated engine. *International Journal of Hydrogen Energy,* 1990; 15:833 - 45.

[39] Air Products and Chemicals I, Safetygram-17. *Dangers of Oxygen-Deficient Atmospheres.* 2009. p. 2.

In: Applied Soft Computing Techniques … ISBN: 978-1-53618-180-7
Editors: Amit Kumar Thakur et al.

Chapter 5

Review on the Performance, Combustion, and Emissions of Butanol-Diesel Mixes in Compression Ignition Engine

Amit Kumar Thakur[1,*], J. V. Muruga Lal Jeyan[1], Ajay Kumar Kaviti[2] and P. S. Ranjit[3]

[1]Department of Mechanical Engineering,
Lovely Professional University,
Phagwara, Punjab, India

[2]Department of Mechanical Engineering,
VNRVJIET, Hyderabad, India

[3]Department of Mechanical Engineering,
Aditya Engineering College (Autonomous),
Surampalem, Andhra Pradesh, India

* Corresponding Author's Email:amitthakur3177@gmail.com.

ABSTRACT

The need for alternative renewable fuels has given rise to considerable interest in addressing the problem of, in particular, higher energy requirements, high oil prices, reduced oil supplies, and ecological emissions of fossil fuels. Alcohols in the form of methanol, ethanol, and butanol have raised sparks. They are the best contenders because of their high oxygen content and high octane content. This comprehensive review aims to highlight the essential characteristics of these alcohols as fuel. Further comparison is carried out with pure diesel to obtain the best results in terms of combustion, performance, and emission parameters for the CI (compression ignition engine). The fuel characteristics study shows that butanol can solve problems with the use of methanol and ethanol as fuel. Progress is also being made in all these alcohols in the production process from different sources. In addition to this, several modus operandi, such as alcohol-diesel fuel mixtures and alcohol-diesel fuel emulsions, alcohol-diesel fuel with nano-particles are discussed and then used in diesel engines. The emissions considered harmful have shown a drastic reduction in the blending of alcohol with diesel fuel and a significant increase in the performance of the diesel engine. Similar studies also show that, due to its improved quality and miscibility with diesel fuel, butanol is a viable alternative to diesel fuel than with methanol and ethanol. In the long run, several key findings and recommendations for future research are essential.

Keywords: alcohols, combustion, performance, emissions, diesel

INTRODUCTION

Due to its excellent fuel consumption and driving capabilities, diesel engines have become an integral part of internal combustion engines in several sectors. Shortly, extensive use of diesel engines will be more prominent. The diesel engine can boast excellent fuel efficiency. However, the central area of concern is the emission of NOx and particulate matter, which needs checking at all levels. Total pollutants and greenhouse emissions of these diesel vehicles are increasing at an alarming rate due to the increasing number of cars. Air pollution has created serious health risks, thus increasing global concern about emissions such as THC, NOx,

SOx, CO, CO_2. Air pollution has drawn the attention of all developing countries to find ways and means to reduce emissions and to implement the necessary measures. Diesel, as a fossil fuel, is depleting at a rapid rate. Thus a substitute fuel is required to meet all the needs of a better environment. The clean energy concept has emerged using various technologies to help reduce the emission of NOx, particulates, and CO_2 and more to fuel burns. It has become imperative to look for different fuel sources, such as gasoline, biodiesel, and diesel esters, to meet the energy consumption requirements and to help control pollution.

Biodiesel, biomethanol, bioethanol, biobutanol, or a mixture of different types of alcohol have the potential to meet the requirements and control the level of air pollution. Studies also reveal the modification of the specification of a diesel engine using these blended fuels to optimize emissions and provide a better environment for each individual. As these alternative fuels are rich in oxygen, adding them to diesel helps to reduce PM emissions that are considered deadly to the environment. Numerous types of research have been conducted on oxygenated additives, such as alcohol and methyl or ethyl esters (biodiesel), to examine the mixing of these additives with diesel fuel. However, other organic compounds, including ethers, glycol ethers, and esters, have been shown to have an effect of oxygen content on fuel properties, combustion, and emissions [1].

The impacts of oxygenated fuels were investigated by Liotta et al., which demonstrated that the use of glycol and ether mixed with oxygen was effective in reducing PM compared to alcohol. Glycol ethers exhibited positive effects, minimizing the emission of smoke, CO, PM, and NOx [2]. With biodiesel, substantial reductions of smoke, PM, and HC emissions are noticeable, while NOx emissions have been higher. Amplified BSFC (break-specific fuel consumption), NOx, carbon monoxide, and CO_2 production occurred when ethylene glycol monoacetate applies to diesel fuels. On the other hand, reduced PM and smoke emissions are observed [3].

Alcohol mixtures have also had a significant impact on finding the best alternative fuel for diesel engines. These mixtures have also contributed to a more substantial reduction in exhaust emissions [4]. Butanol has emerged as one of the potential alternative diesel fuels. It has produced excellent results in diesel engines in varying proportions [5]. n-Butanol is a primary form of alcohol that is a useful alternative fuel mixed with methanol and ethanol in the IC engine. The majority of the n-Butanol fuel attributes are similar to those of diesel fuel [6,7]. Due to their comparable fuel performance with traditional diesel fuel, n-Butanol will reduce global fuel consumption [8–11]. Although the importance and efficiency of alcohol-diesel combustion are known, further research is needed to improve performance. It is also required to ensure the preservation of the engines used in such mixtures and build appropriate distribution systems.

For this reason, extensive studies to assess the performance and emission properties of diesel engines with different alcohol-diesel combinations are needed. The increasing interest in diesel additives becomes essential, as several studies in this context between 2005 and 2019 reveal this. Between 2005 and 2019, more than two hundred articles came, and several studies justify the main concern for diesel-alcohol blends. These articles focus on various factors for the next generation of alcohol-diesel fuel. These include the origins of ethanol, the behavior of phases and the characteristics of alcohol-diesel mixtures, the influence of alcohol-diesel fuel mixtures on perforation and diesel emissions, and the effects of timing of injections and pressure on the exhaust emissions of diesel fuel engines. The main aim of this research is to undertake a broad review of more than 14 years of literature on the use of diesel-butanol fuel in diesel engines and its impact on combustion, productive consumption, and emissions.

REASONS FOR ADVOCATING BUTANOL

Alcohol was known to be the fuel for internal combustion engines; however, testing and developmental study needs more research to promote the combustion cycle. Alcohols have developed themselves as a useful alternative for the substitution of fossil fuels, which are depleting every day. The usage of alcohol has prompted numerous experts to talk about increasing fuel performance and rising toxic exhaust pollution. The essential physicochemical properties of butanol discussed below. Butanol is an alcohol with four carbon atoms, it has four isomers, and all isomers have specific processing methods, but they have the same ability. They also have various solubilities, but they have similar molecular weights and functional groups. Linear alcoholic liquor is n-butanol. It is a colorless, flammable substance with a slightly hydrophobic flavor of banana and a strong alcoholic smell. Visual exposure may contribute to inflammation of the eyes and the scalp, and their vapors have a narcotic impact on the mucous membranes at elevated doses. It is miscible, as are the most popular organic solvents. Iso-Butanol is like a mildly musty scent, a colorless liquid. Whereas2-butanol is a flammable, colorless liquid with synthetic polar solvents that is partly water-soluble and fully miscible. t-Butanol is a solvent that is translucent and is highly water-soluble and miscible for ethanol and diethyl ether. It tends to be stable at room temperature (the melting point just above 251C). Section 1 lists the essential properties of various butanol isomers[1].

COMPARISON OF PHYSICOCHEMICAL PROPERTIES

Table 2 shows the physicochemical properties of butanol and their contrast with gasoline, methanol, and diesel fuel. It is symptomatic that because of lower hygroscopicity, butanol has emerged as a definite contender in the area of renewable energy as a possible fuel. Climate shifts had little effect on the physicochemical properties of butanol. Butanol wins over gasoline and methanol because of the higher flash point benefit. Similar to lower carbon chain alcohols, butanol has the upper hand in power production and will undoubtedly produce higher mileage. The substantial octane degree is one of the hallmarks of butanol.

Table 1. Comparison of various physicochemical properties of different isomers of butanol [1]

Property	n-butanol	iso-butanol	2-butanol	3-butanol
Density at $20^{0}C(g/cm^{3})$	0.810	0.802	0.806	0.781
Boiling point (^{0}C)	118	108	99	82
Water solubility (g/100 mL)	7.7	8.0	12.5	Miscible
Flashpoint (^{0}C)	35	28	24	11

Table 2. Comparison of various physicochemical properties of methanol, ethanol, and butanol with those of diesel fuels [1]

Property	Methanol	Ethanol	Butanol	Diesel fuel
Density at 200C (g/ml)	0.7866	0.7851	0.8098	0.829
Boiling point, (0C)	65	78	117–118	187–343
Motor octane no	104	102	78	–
Research octane no	136	129	96	15–25
Cetane no	3.8	5–8	E25	40–55
The energy density (MJ/L)	16	19.6	27–29	46
Energy content (Lower heating value) (MJ/kg)	20.1	26.9	33.1	42.8
Higher heating value (MJ/kg)	22.7	29.7	36.6	44.8
The heat of vaporization (MJ/kg)	1.2	0.92	0.43	0.23–0.60
Freezing point, (0C)	97	114	89.3	40–34
Flashpoint (0C)	12	13	35	74
Autoignition temp (0C)	463	423	397	235
Flammability limits, volume (%)	36	3.3–19	1.4–11.2	6
Solubility in 100 g of water	Miscible	Miscible	Immiscible	Immiscible

By comparison to ethanol and methanol, butanol has a longer hydrocarbon line, rendering it non-polar and close to petrol. Thanks to its miscibility with petrol, butanol has been seen to be an efficient fuel that can be quickly substituted either partly or entirely by gasoline. Methanol and ethanol must be combined with diesel to achieve the required fuel for service. Butanol may be used directly either in a limited proportion or in a vast quantity to power the engine. The lower chain of alcohol is very corrosive, and this issue is taken care of by butanol. Viscosity relies on the duration of the carbon chain, and in this sense, butanol has a higher kinematic viscosity. Butanol mixes with fuel at any rate. It cannot be separated when combining it with water, even at a lower temperature. Butanol continues to reduce the degree of solidification of diesel fuel mixtures at low temperatures.

Nevertheless, the main issue with butanol used as a diesel mixing element is the low CN, which may limit the capacity of the mixture to auto-ignite while combustion is involved. Much work is required to develop butanol as a possible fuel globally. Action is needed to find a few suitable additives or CN enhancers that can help increase its fuel properties for improved fuel use. It should also boost the efficiency parameters for diesel engines with lower emissions [1].

THE OUTCOME OF DIESEL-BUTANOL MIXES ON PERFORMANCE AND EMISSIONS OF DIESEL ENGINE

The diesel engine is the most potent control generator among all recognized internal combustion engine types. Large trucks, small engines, and building equipment are powered by diesel engines worldwide. The introduction of diesel cars with high-volume combustion engines contribute to dominate power systems with potential automobiles. Nevertheless, further change is required before this occurs, as global environmental regulations, coupled with pollution requirements, are getting further stringent [1]. Various techniques are there to reduce the pollution of diesel engines. One of the critical means of regulating the emissions of diesel engines was adding alcohol with diesel fuel to provide extra oxygen during the combustion cycle. Dual-fuel gasoline diesel systems come in a variety of forms. High autonomy of diesel fuel guarantees the ignition of ethanol in dual-fuel service. Numerous CI engines have been used with alcohol-diesel blends to monitor engine output and exhaust emissions [12]. Blending alcohol with diesel fuel can help to increase the efficiency of diesel engines and, in turn, can reduce the release of toxic gasses from the vehicle exhaust.

PROGRESS IN PERFORMANCE, COMBUSTION, AND EMISSIONS OF BUTANOL-DIESEL FUEL MIXES

n-Butanol has an enormous ability to be the safest fuel additive to be found in a diesel engine. Butanol has emerged as the next-generation fuel due to its modest volume of cetane. It has hardly ever been investigated in diesel engines [13]. Butanol has created a great deal of interest to all researchers employed in the field of green energy because its properties are far more like diesel fuel [14]. The U.S. regulators have licensed biobutanol. Environmental Protection Agency should be mixed as an oxygenate with gas. Nonetheless, the future is promising for butanol, which will appear as one of the better alternatives for fossil fuel. There is no research on the study of butanol-diesel fuel mixtures as a replacement fuel for the monitoring of efficiency criteria and the release of dangerous gases [15–17]. Miers et al. examined the durability and pollution of the petrol-powered Mercedes-Benz C220 diesel vehicle (20 percent and 40 percent by volume)[15]. The findings of the analysis offered higher hydrocarbon and carbon monoxide values as the volume of butanol decreased. 40% of butanol significantly improved the amount of NOx pollution. Rakopoulos et al. [16] studied the output and emissions of a diesel engine driven at a speed of 2000 rpm under various loads with the aid of n-butanol combinations. The findings of the experiment revealed that NOx and CO pollution had decreased as low contact with smoke intensity. By comparison, the release of HC increased with the usage of butanol-diesel blends. There was a small improvement in SFCs, as the engine's BTE rose marginally. Yao et al.[18] carried out their tests using diesel and n-butanol on a heavy-duty diesel generator. He modified the amount of exhaust gas distribution to ensure that NOx emissions stayed stable. The trial is within limits as soot, and CO pollution is small without impacting the emission of BSFC and NOx. The secretion of HC increased with a slight rise in BTE. Al-Hasan et al.[19] tested the output of diesel engines at various engine speeds with specific iso-butanol mixtures. The exhaust temperature, BP, and BSFC increased, and the BTE and air-fuel ratio decreased with an improvement in engine rpm. Thirty percent of iso-butanol offered satisfactory values in terms of engine efficiency. However, a further 40 percent resulted in substantial variations in engine output. Karabektas et al. checked the engine output and exhaust discharge [20] at variable volume and various speed in utilizing iso-butanol diesel fuel formulations. With 10% isobutanol, optimum BTE value results. By using iso-butanol-diesel fuel mixes, carbon monoxide and NOx emissions have reduced with higher HC emissions. The addition of the iso-butanol amount in the mixture of fuel resulted in lower values of engine speed, BTE, and EGT.

In this study[21], the contrast of pure diesel with iso-butanol-diesel resulted in a higher smoke intensity and a decrease in smoke volume when iso-butanol combines with diesel. The introduction of 10 percent of isobutanol oil to olive oil by 5-10 percent has helped to increased pollution of smoke, CO, and HC. Throughout their theoretical study, Ozsezen et al. [22] investigated the effect of diesel fuel mixtures on the efficiency and combustion of iso-butanol (5 percent, 10 percent, and 15 percent). BSFC strengthened with 5 percent and 10 percent isobutanol and displayed the highest prices by 15 percent. BTE decreased with the isobutanol content. Peak gas pressure and heat release rate increased significantly, although smoke emission, NOx, and carbon monoxide decreased. There was, however, an improvement in the amount of HC pollution. In his experimental research, Biao et al.[23] reported a small increase in overall brake and brake torque using 30 percent butanol when tests are done with butanol (0 percent, 20 percent, and 30 percent)-diesel fuel mixtures. Through the usage of 30% butanol fuel blends, soot emissions reduced significantly relative to 20% butanol blends. Also, Dogan [24] studied diesel fuel production and exhaust at constant rpm. BSFC and BTE were slightly rising in combustion with increased n-butanol levels, with specific engine loads of n-butanol (5 percent, 10 percent, 15 percent, and 20 percent)-diesel fuel. However, the levels of HC grew in no small point. Rakopoulos [25] performed an inquiry into the combustion of diesel engines with n-butanol (8, 16, and 24 percent by volume)-diesel fuel mixtures. The ignition lag decreases as the analyst find gaps in the fuel ignition pressure diagrams, the decreased smoke intensity, and release of NOx associates with improved demand for BTE. Rakopoulos et al. further explored the combustion of the six-cylinder DI engine fuelled by ethanol-diesel and n-butanol-diesel fuel mixtures [26]. Ethanol fuel mixtures decreased the smoke level when the blend with n-butanol gasoline boosted the NOx pollution by more than ethanol mixtures. Compared to fossil fuel, the BSFC and BTE levels increased for the n-butanol combinations. The ignition interval decreased by a steady reduction in the average cylinder load. During the first phase of the combustion cycle, the temperature of the flame has declined significantly [27].

In a study conducted by Ozdor [28], an improvement in power production with 10% of the same volume of fuel could contribute to the removal of cycle-to-cycle volatility. Such research investigated broad pressure differences for diesel fuel engines utilizing stochastic techniques under various loads and injection timings. The integration of butanol into diesel fuel would reduce CN butanol and may contribute to problems with cyclic volatility. Rakopoulos et al. [25] studied the combustion of diesel-n-butanol mixtures and cyclic instability in the DI diesel engine. The pattern of cyclic variation is non-systematic. An analysis of the fuel pressure pipeline for delays in ignition found that the form of fuel pump device used did not affect the amplitude of the cyclic load. It also shows that n-butanol does not affect efficiency or emissions. Yao et al. [29] investigated the effect of EGR on the ability and pollution of a high-speed diesel power plant using specific burning criteria using butanol in 10% and 20% by volume combined with diesel fuel. Medium EGR yielded reliable results when compared with high and low EGR ratios. Soot and NOx drop to a lower value. Tornatore et al. studied the combustion results of the compression ignition engine [30] with 80% butanol and 20% diesel fuel. Tests on the combustion cycle for corresponding improvements in the timing of injections and EGR levels at a set injection environment. Mixtures used demonstrated improved combustion time, reduced smoke, and NOx pollution. B20 blend with 50% EGR is most suitable for decreased engine emissions at the time of late injection. High rise in-cylinder gas pressure, HRR was observed with a pause in ignition using butanol or iso-butanol in the rising volume of diesel. Thermal output increased with butanol-diesel mixtures from the previous rating. In another research, Yosimoto et al. [31, 32] examined the usage of oxygenated organic compounds (alcohols) in rapeseed oil.

Compared to rapeseed oil,1-propanolor1-butanol resulted in proper combustion with decreased smoke production, little reduction in NOx production, and a rise in BSFC efficiency. Butanol blends, diesel fuel, and croton oil are the fuel to test and validate the efficiency, emission, and combustion characteristics of the Lujaji et al. CI engine [33]. The integration of butanol and diesel fuel into vegetable oil has demonstrated improved fuel properties with low values of BSFC and BTE. The right combination of butanol and croton oil and diesel mixtures resulted in an improvement in the HRR at a higher cylinder pressure size. For the butanol blend, the levels of CO_2 and smoke substantially decreased, and the NOx was similar to that of the processed diesel fuel used. In their experimental research, Altun et al. [34] used a mixture of n-butanol-diesel-biodiesel to measure the performance of the CI motor. With the addition of n-butanol in the diesel-biodiesel (B20) combination, BSFC and BTE levels were significantly higher than those of diesel-biodiesel (B20). CO and HC indicated low value, and NOx emissions were comparable at small engine loads and increased at high engine loads. Smoke visibility shows a substantial decrease over the entire engine test spectrum. It was operating with turbocharged diesel motors, Rakopoulos et al. [35] investigated the effects of applying biodiesel and n-butanol to diesel fuel. Remarkable findings have shown that the latency of the turbocharger was the cause of the pollutant pollution in both blends. Biodiesel blend has been held accountable for irregular combustion and volatility. Smoke emission decreased for the mix of bio-diesel and with a mixture of n-butanol. NOx levels rose with all combinations as compared to pure diesel fuel. Three blends (30 percent biodiesel, 25 percent n-butanol, and diesel oil) are of great importance to study the processes of development of smoke and NOx with varying engine speed. Results displayed that smoke emissions are decreasing further when 25 percent of n-butanol combines with fuel[36]. NOx emissions have been substantial relative to conventional diesel, and the n-butanol mixtures have reduced NOx. Smoke emission decreases for n-butanol combinations than in biodiesel mixtures. LIU et al. [37] studied combustion and soot production characteristics at specific temperatures of pure butanol and pure soya biodiesel at varying concentrations of oxygen (21 percent, 16 percent, and 10.5 percent). The effect was higher uniform peak pressure, as butanol has a higher propensity for thermal performance. The combustion strength and luminosity of butanol are higher than soya. The total soot, though, decreased as the temperature for both fuels grew. Jin et al. [38] have recorded natural luminosity and soot emissions of 1200 K by using Butanol as an addition to the petrol additive.

The usage of pentanol-diesel in the 5, 10, and 20 percent range tests were carried out on a DI diesel engine to determine thermophysical properties, combustion properties, engine efficiency, and engine emissions. Results show that the calorific content, density, and kinematic viscosity of PE 20 is minimal. The in-cylinder pressure of both blends is minimum at 25% of the engine load utilizing the mix. The average heat release levels of PE5, PE10, and PE20 are low, and the HRR of diesel is high. However, the BSFC of PE5 produced improved performance, and the BTE also through. Throughout the case of PE 20, CO and CO_2 emissions decreased [39].

In this study, the researcher suggested a novel concept of creating a new reduced chemical kinetic method by combining n-heptane-n-butylbenzene-n-polycyclic aromatic hydrocarbons in differing amounts. This experiment plans to track engine results with ignition delay, laminar flame intensity, and different emissions and overall combustion behavior. The findings indicated that the experimental values are almost consistent with the expected combustion rates and emissions of soot, NOx, and CO, validating the current diesel engine simulation procedure. Similar to mere gasoline, lower rates of carbon and NOx resulted in n-pentanol/gasoline mixtures. The ignition period was longer, and thus adequate combustion of the mixture was ensured[40].

To equate toxicity, performance, and combustion properties with those of pure diesel and CalophyllumInophyllum methyl ester, ternary mixtures of n-hexanol, diesel, and CalophyllumInophyllum methyl ester have experimented. Hexanol mixtures display a higher BTE and a lower BSFC than those of CalophyllumInophyllum methyl ester. Hydrocarbon, smoke intensity, and CO are smaller than fossil fuel. The findings gave a strong example of potential solutions to pure diesel, and a proposal displays the use of 30 to 40 percent n-hexanol fusion effectively rather than pure diesel. The production of NOx and CO_2 was marginally large relative to pure diesel[41]. The efficiency, emission, and combustion characteristics of diesel fuel blends with tomato seed- oil methyl ester (TSME), n-butanol, were studied at different loads in these experiments. TSME blends with n-butanol diesel at various levels of 5, 10, and 15 percent at a steady state of 10%. BTE, EGT, and NOx rise to a higher value. The emission of hydrocarbons, carbon monoxide, and smoke minimizes. High pressure and HRR were seen on the top line, offering improved engine characteristics by combining tomato seed oil for n-butanol diesel [42].

High performance at specific engine loads and two exhaust gas circulation (EGR) (15 percent and 30 percent) concentrations for n-pentanol, Dimethyl Ether (DME), and diesel mixtures becomes the objective of this study. The three research mixtures, D85DM15, D65P35, and D60DM20P20, were applied to the specific amounts of DME and n-pentanol in baseline diesel. Particulate matter (PM) is least for D60DM20P20 relative to healthy diesel. The D60DM20P20 blend produced the lowest emissions of NOx and PM. The THC and carbon monoxide output of oxygenated blends were higher when measured against gasoline. Overall, the net result indicates a more robust ability for the mixture of n-pentanol, DME, and gasoline (D60DM20P20). These dangerous pollutants have increased with the introduction of EGR[43].

In this study, comparative research with diesel and 15% butanol (Bu15) using the CFD CONVERGE application gets into action. The upshot found that the single-injection process decreased NOx, carbon monoxide, and soot emissions. In contrast, UBHC emissions rose with the introduction of butanol. The heat release rate (HRR) fell in the pilot injection technique for the first fuel but improved marginally for the second fuel injected. The minimum ISFC value of 10% of the pilot injection fuel was recorded [44]. Pentanol and Hexanol were combined to form Pen30 and Hex30 with an amount of 30% diesel for use in diesel engines to attain the best engine efficiency. In light of diesel, the combustion and output characteristics become the objective of the study. The analysis found that the Pen30 combination had more ignition times, higher temperatures, and more significant heat release (HRR) than the Hex30 combination. These configurations also increased thermal efficiency, at the expense of fuel usage. Pen30 displays a higher value of BTE and low smoke densities. NOx emission drops for Hex30, while HC emission for diesel fuel is minimum. CO emissions were also low for Pen30, rendering it the perfect replacement for diesel with limited engine model changes [45].

The compression ignition (CI) engine uses 10, 20, and 30 percent clean diesel and butanol/diesel mixture to obtain the necessary output results, combustion, gas emissions, and nano-particle emissions. The use of butanol in diesel decreased the overall accumulation of particulate matter [46]. This research uses Graphene oxide (GO), graphenenanoplatelets (GNP), and multi-walled carbon nanotube. The aim is to examine the efficiency and release characteristics of the diesel engine using Jatropha methyl ester 60 percent (in volume) and n-butanol fuel 40 percent (in volume) (JME40B). The elevated pressures and BSFCs of the JME40B blended fuel reduce much in comparison to pure diesel fuel. The development of JMB40B carbon nanomaterials significantly decreased direct fuel combustion and reduces NOx, carbon monoxide, and unburned hydrocarbons. As a consequence of the use of carbon nanomaterials by combining a robust n-butanol ratio with jatropha biodiesel gasoline, ultra-low NOx, CO, and UHC emissions result with a significant degree of thermal efficiency [47].

In this experiment, butanol and diethyl ether (DEE) combines with plastic waste pyrolysis fuel in the C.I. engine to measure efficiency and pollution. The impact of the application of castor oil on such combustion fuel is also studied. For testing, the engine uses four new fuels with the same oxygen content. The findings reveal that the application of DEE to plastic oil waste produces more HC and smoke pollution than the addition of butanol under low engine operating conditions. HC and smoke reduce to a lower value. DEE and castor oil blends also contributed to a reduction in emissions of HC, CO, NOx, and smoke [48].

This research underscored the efficacy of combustion and the physical and chemical properties of DI diesel engine soot particles. At 15% and 30% by amount, the fuel used was n-pentanol combined with biodiesel. Delayed start and shortened combustion of pentanol-diesel mixtures become noticeable. The amount of the PM(particulate matter) declined along with the reduced amount of the essential carbon (EC) emission of biodiesel. PM further decreases when pentanol combines with biodiesel.

In contrast, organic carbon (OC) emissions rise to a higher amount for both biodiesel and gasoline. Higher amounts of pentanol in blended fuel resulted in lower overall PAH particulate-phase emissions[49]. This study aimed to analyze the efficiency of butanol-microalgae biodiesel-diesel mixtures. Both fuel blends display higher rates of BSFC and BTE than regular petroleum fuel. There was an improvement in NOx production, as 2 percent of microalgae mixes with diesel blends. With the usage of butanol and water, NOx emission reduces as compared to fossil fuel. PM reduces for both the mixtures. Interestingly, just 2% of microalgae biodiesel reported a decrease in CO emissions and decreased CO emissions when diesel blends were used [50].

This research mixed diesel with biodiesel and biodiesel-vegetable oil-alcohol (propanol and pentanol). Research fuel mixtures are formulated and tested on a diesel engine using a splash mixing technique. According to the engine samples, BSFCs of diesel-biodiesel-vegetable oil-propanol improved compared to diesel-biodiesel and diesel-biodiesel-vegetable oil-pentanol at all engine loads. Diesel-biodiesel-vegetable oil-propanol obtained a lower NOx pollution compared to the diesel-biodiesel mix.

Carbon monoxide and hydrocarbon emissions have risen by applying increasing of the higher amounts of alcohol to diesel-biodiesel [51].

The research looked at the fuel characteristics, engine efficiency, and emission characteristics of diesel fuel, fake flax biodiesel, and its butanol mixtures. The torque loss was observed based on the expected experimental values of 10% alcohol in diesel and biodiesel fuels. The torque benefit was also reduced by an improvement in the butanol ratio, resulting in a decrease in power values. By adding butanol, SFCs have increased with lower CO and CO_2 levels with the addition of alcohol in diesel [52]. To test the combustion properties of the fuel to be used in commercial aircraft CI engines, the investigator used diesel and Chinese aviation kerosene (RP-3) along with pentanol in differing amounts. Combustion performance of the mix using 60 percent kerosene and 40 percent pentanol improved along with thermal quality also produced better results. The mixture of kerosene and pentanol retarded ignition timings, however, these mixtures demonstrated a lower value of actual fuel usage, which was less than what used to come from jet fuel. Make slight changes to the aircraft diesel engine, kerosene-pentanol mixtures can soon be used [53]. Waste cooking oil biodiesel was combined with n-pentanol to test the results of combustion and spray ignition.

The rise of the atmospheric temperature to the defined amount led to the reduction of the liquid length. Thereby the slope of the n-pentanol mixtures increases. The added advantage of such mixtures demonstrated a strong propensity for soot oxidation [54]. The investigator notices a reduction of PM emissions with a small improvement in physicochemical properties utilizing ultra-low sulfur diesel in the standard type of mixtures of ethanol and n-butanol [55]. The solubility test for ethanol-diesel combinations initiates with the corresponding introduction of n-butanol in 6 to 10 percent volume in this experimental analysis. The energy content, along with the cetane level, viscosity, and density, increases to a high value when n-butanol mixes with ethanol-diesel mixtures [56].

The primary objective of the study was to investigate the effect of the importance of EGR on emission parameters using ethanol and butanol as fuel. The results showed higher rates of carbon monoxide and smoke emissions under different test conditions with the implementation of EGR, whereas HC emission displays the lower value. The usage of butanol as fuel resulted in a loss of combustion efficiency [57]. Ethanol, n-butanol, and isopropyl alcohol are mixed with pure diesel to evaluate the efficiency and pollution values of the CI motor. The usage of pure diesel with 15% ethanol and 30% isopropyl alcohol improved the brake thermal output of the engine with a small decrease in BSFC. Use ethanol and n-butanol at the same level as 30 percent of isopropyl alcohol created fewer carbon dioxide emissions. At a comparable volumetric mix of these three oils, smoke visibility and NOx pollution are high. Better thermal performance and lower pollution are noticeable, using 70% diesel and 30% isopropyl alcohol [58].

Mixtures of n-butanol-diesel and ethanol-diesel were used by the researcher to experiment to explore cycle-by-cycle variability, output, and exhaust pollution in a diesel engine. In comparison with pure diesel, these two mixtures achieved improved fuel output and brake-specific fuel consumption values. CO and NOx levels have reduced with n-butanol and ethanol in diesel fuel. Cycle-by-cycle variability shows a growing pattern with a resulting rise in low-speed alcohol-diesel mixtures [59]. Different loads of engines determine the oxidation and particulate pollution of the biodiesel-pentanol (BP) fuel system. The start of combustion was delayed with decreased pentanol values when all BP mixtures produced faster burning. Brake Thermal Performance has improved with all BP blends. Particulate matter emissions are low with the introduction of pentanol in biodiesel. This research indicates that pentanol-biodiesel can also be known to be a potential fuel for diesel engines [60].

STUDY REVIEW AND FUTURE RECOMMENDATIONS

Available studies and analysis indicate that emissions from smoke, NOx, and carbon dioxide have declined dramatically. In contrast, emissions of HC have increased due to butanol-diesel mixtures related to proper diesel fuel. Several forms of study, on the other hand, reported the emission of fewer hydrocarbons. Nonetheless, other studies demonstrate that butanol diesel mixtures will reduce soot pollution. However, NOx emission volatility depends on butanol quality and engine conditions. Various analytical methodologies and engine modifications are noticeable with these differences. Butanol or isobutanol in an increasing trend gives a lower value of BTE. Consumption of diesel-butanol or iso-butanol can be increase with turbocharged or low-heat-release engines [17]. Butanol provided improved mixing power and BSFC. However, with butanol application, a decrease in carbon monoxide and HC release for diesel-biodiesel mixtures with a decreased NOx pollution penalty is observed.

n-butanol thus provides better thermal efficiency, pollution, and higher miscibility of CN and diesel fuel than lower alcohol levels. Compared to diesel fuel, N-butanol has a lower CN as the biggest drawback. Low CNs may be called upon to boost auto-ignition and precipitate combustion delays, thus increasing soot emissions for potential diesel combustion processes. The increase in injection pressure and timings helps to analyze the efficiency and pollution of the diesel engine. However, detailed research contributes to better results in the immediate future. EGR is known to be the most crucial parameter for regulating the combustion, efficiency, and pollution of diesel engines, in particular NOx and soot. Some EGR-related study reveals good results, but potential science also requires more comprehensive studies to produce better outcomes. To increase pollution and engine performance, fuel injection technologies such as fumigation and dual fuel injection need to be more complicated with diesel combustion [1].

Conclusion and Future Recommendations

The need for the hour to support and preserve humanity and the world is to substitute the fossil fuel supply with an accessible and renewable tool. Dependency on fossil fuels has devastated the state of life, and suffering persists as global deposits deplete day by day. Fossil fuels produce an immense volume of greenhouse gas pollution that greatly outweigh environmental requirements. It is necessary to pursue alternate energies, which would have the ability to substitute diesel fuel, which should ensure maximum protection in the future for all environmental issues. Under this respect, the higher chain of alcohol has demonstrated the ability to be the most significant substitute source of fuel to rising greenhouse gas emissions. Ethanol and methanol have obtained excellent performance, but butanol can solve the issues associated with the usage of methanol and ethanol. Butanol is a high-quality fuel using technological advanced biotechnological processes and has comparable fuel content. It may either be combined with fuel or used in a streamlined shape in a diesel engine without any modification to the motor. Owing to the healthy oxygen level and lower sulfur and carbon level, butanol has improved from industrial diesel fuel in terms of lower pollution. Butanol will substantially eliminate fossil fuel usage and could quickly become the primary source of energy for compression ignition (CI) engines. The research so far is beneficial in classifying butanol as the better for DI engines.

However, there is more scope of conducting extensive investigations to test the impact of butanol-diesel mixtures on the efficiency, emission, and combustion activities of diesel engines. The exploitation of cellulose biomass at a reduced cost to further improve the use of biofuels as a renewable fuel can be demonstrated more effectively through a study of policy reforms and the advancement of emerging technology. Butanol has developed itself as the most effective additive to diesel fuel due to its higher value of cetane number, low vapor pressure, and lower miscibility. Butanol is known to be less corrosive and has a higher volume of strength. Butanol is the safest addition to hydrocarbon oils.

Owing to its intense energy density and weak hygroscopicity, the higher butanol iso-butanol,1-butanol, 2-methyl-1-butanol, 3-methyl-1-butanol, and the higher butanol chain has an immense ability relative to lower alcohols. For improved usage of diesel engines, extensive work on these alcohol chains gives better results. Throughout this throughout-depth analysis, smoke, carbon monoxide, and THC release decrease with higher

NOx emissions, the amount of butanol in the fuel mixture increases. Careful analysis culminated in elevated levels of BTE and BSFC for n-butanol, isobutanol, and n-butanol-diesel blends and lower exhaust gas temperatures. In contrast, iso-butanol-diesel hybrid combustions saw an improvement in BSFC, with lower engine performance, BTE, and exhaust gas temperature. Improved pressure injection results in a higher value of NOx and decreased smoke visibility. Injection timings often influence the output of the diesel engine. Experimental areas and comprehensive work establish modern injection technologies to boost the efficiency of diesel engines and reduce exhaust pollution.

Eventually, butanol-diesel fuel mixtures boosted engine efficiency without having any significant impact on engine performance. Fuel additives and cetane number, increase fuel characteristics, and the stability of mixtures. Further analysis and experimental work should be conducted, in particular, for butanol-diesel combinations, for the evaluation of stability. Verification of the function of various additives, which have a better perspective impact on butanol-diesel fuel mixtures as next-generation fuel, is vital.

REFERENCES

[1] Kumar S, Cho J H, Park J, Moon I. Advances in diesel–alcohol blends and their effects on the performance and emissions of diesel engines. *Renewable and Sustainable Energy Reviews* 2013; 22: 46–72.

[2] Liotta FJ, Montalvio DM. *The effect of oxygenated fuels on emissions from a modern heavy-duty diesel engine*. SAE technical paper no. 932734; 1993.

[3] Lin CY, Huang JC. An oxygenating additive for improving the performance and emission characteristics of marine diesel engines. *Ocean Engineering* 2003; 30:1699–715.

[4] Wang LJ, Song RZ, Zou HB, Liu SH, Zhou LB. Study on combustion characteristics of methanol–diesel fuel compression ignition engine. *Proceeding of Institution of Mechanical Engineers Part D: Journal of Automobile Engineering* 2008; 222:619–27.

[5] Brekke K. *Butanol: an energy alternative*. /http://www.ethanol.org/pdf/contentmgmt/March_07_ET_secondary.pdfS.

[6] De Freitas LD, Kaneko S. Ethanol demand under the flex-fuel technology regime in Brazil. *Energy Economics* 2011; 33:1146–54.

[7] Ajav EA, Akingbehin OA. *A study of some fuel properties of local ethanol blended with diesel fuel*. Ibadan, Nigeria: Department of agricultural engineering, Faculty of Technology, University of Ibadan; 2002. http://ecommons.library.cornell.edu/bitstream/1813/10277/1/EE%2001%20003.pdfS.

[8] Agarwal, AK. Biofuels (alcohol and biodiesel) applications as fuels for internal combustion engines. *Progress in Energy and Combustion Science* 2007; 33:233–71.

[9] Kremer FG, Jordim JLF, Maia DM. *Effect of alcohol composition on gasoline vehicle emissions*. SAE paper no. 962094; 1996.

[10] Di Y, Cheung CS, Huang Z. Experimental study on particulate emission of a diesel engine fueled with blended ethanol–dodecanol–diesel. *Journal of Aerosol Science* 2009; 40(2):101–12.

[11] Bayraktar H. An experimental study on the performance parameters of an experimental CI engine fueled with diesel–methanol–dodecanol blends. *Fuel* 2008; 87(2):158–64.

[12] Chotwichien A, Luengnaruemitchai A, Jai-In S. Utilization of palm oil alkyl esters as an additive in ethanol–diesel and butanol–diesel blends. *Fuel* 2009; 88:1618–24.

[13] Harvey BG, Meylemans HA. The role of butanol in the development of sustainable fuel technologies. *Journal of Chemical Technology and Biotechnology* 2011; 86:2–9.

[14] Szwaja S, Naber JD. Combustion of n-butanol in a spark-ignition IC engine. *Fuel* 2010; 89(7):1573–82.

[15] Miers SA, Carlson RW, McConnell SS, Ng HK, Wallner T, Esper JL. *Drive cycle analysis of butanol/diesel blends in a light-duty vehicle.* SAE paper no. 2008-01-2381; 2008.

[16] Rakopoulos DC, Rakopoulos CD, Giakoumis EG, Dimaratos AM, Kyritsis DC. Effects of butanol–diesel fuel blends on the performance and emissions of a high-speed DI diesel engine. *Energy Conversion and Management* 2010;51(10):1989-97.

[17] Rakopoulos DC, Rakopoulos DC, Hountalas DT, Kakaras EC, Giakoumis EG, Papagiannakis RG. Investigation of the performance and emissions of bus engines operating on butanol/diesel fuel blends. *Fuel* 2010; 89(10):2781–90.

[18] Yao M, Wang H, Zheng Z, Yue Y. Experimental study of in-butanol additive and multi-injection on HD diesel engine performance and emissions. *Fuel* 2010; 89(9):2191–201.

[19] Hasan MI, Momany M. The effect of iso-butanol–diesel blends on engine performance. *Transport* 2008; 23(4):306–10.

[20] Karabektas M, Hosoz M. Performance and emission characteristics of a diesel engine using iso-butanol–diesel fuel blends. *Renewable Energy* 2009; 34(6):1554–9.

[21] Asfar KR, Al-Rabadi TH fuel blends in compression ignition engines, American Institute of Aeronautics and Astronautics (AIAA 2003-6027). In: *Proceedings of the 1st International energy conversion engineering conference,* 17–21 August 2003, 432 Portsmouth, Virginia, USA. p. 433.

[22] Ozsezen AN, Turkcan A, Sayin C, Canakci M. Comparison of performance and combustion parameters in a heavy-duty diesel engine fueled with iso-butanol/diesel fuel blends. *Energy Exploration and Exploitation* 2011; 29: 525–41.

[23] Biao D, Zhiyu H, Zheng C, Yun L, Xiaokang Z, Guoguang W. A study on performance and soot emissions of a high-speed diesel engine fueled with the butanol-diesel blend. In *Proceedings of the international conference on electric information and control engineering*, ICE 2011, Article no. 5777155, p.4742–45.

[24] Dogan O. The influence of n-butanol/diesel fuel blends utilization on a small diesel engine performance and emissions. *Fuel* 2011; 90(7):2467–72.

[25] Rakopoulos CD, Rakopoulos DC, Giakoumis EG, Kyritsis DC. The combustion of n-butanol/diesel fuel blends and its cyclic variability in a direct injection diesel engine. *Proceeding of Institute of Mechanical Engineers Part A: Journal of Power and Energy* 2011; 225:289–308.

[26] Rakopoulos DC, Rakopoulos CD, Papagiannakis RG, Kyritsis DC. Combustion heat release analysis of ethanol or n-butanol diesel fuel blends in heavy-duty DI diesel engine. *Fuel* 2011; 90:1855–67.

[27] Ozdo N, Heywood J. *Internal combustion engine fundamentals*. New-York, NY, USA: McGraw-Hill; 1988.

[28] Ozdor N, Dulger M, Sher E. *Cyclic variability in spark-ignited engines, a literature survey*. SAE paper no. 950683, 1994.

[29] Yao M, Zhang Q, Liu H, Zheng Z, Zhang P. *Diesel engine combustion control: medium or heavy EGR*. SAE paper no. 2010-01-1125.

[30] Tornatore C, Marchitto L, Mazzei A, Valentine G, Corcione FE, Merola SS. Effect of butanol blend on the in-cylinder combustion process, part 2: Compression ignition engine. *Journal of KONES Powertrain and Transport* 2011; 18(2): 473–83.

[31] Yosimoto Y, Onodera M, Tamaki H. *Performance and emission characteristics of a diesel engine fueled by vegetable oils*. SAE paper no. 2001-01-1807/4227.

[32] Yosimoto Y, Onodera M. *Performance of a diesel engine fueled by rapeseed oil blended with oxygenated organic compounds*. SAE paper no. 2002-01-2854.

[33] Lujaji F, Kristo´f L, Bereczky A, Mbarawa M. Experimental investigation of fuel properties, engine performance, combustion, and emissions of blends containing croton oil, butanol, and diesel on a CI engine. *Fuel* 2011; 90(2):505–10.

[34] Altun S, Oner C, Yasar F, Adin H. Effect of n-butanol blending with a blend of diesel and biodiesel on performance and exhaust emissions

of a diesel engine. *Industrial and Engineering Chemistry Research* 2011; 50: 9425–30.

[35] Rakopoulos CD, Dimaratos AM, Giakoumis EG, Rakopoulos DC. Study of turbocharged diesel engine operation, pollutant emissions, and combustion noise radiation during starting with biodiesel or n-butanol diesel fuel blends. *Applied Energy* 2011; 88:3905–16.

[36] Rakopoulos CD, Dimaratos AM, Giakoumis EG, Rakopoulos DC. Investigating the emissions during the acceleration of a turbocharged diesel engine operating with biodiesel or n-butanol diesel fuel blends. *Energy* 2010; 35:5173–84.

[37] Liu H, Lee CF, Huo M, Yao M. Combustion characteristics and soot distributions of neat butanol and neat soybean biodiesel. *Energy and Fuel* 2011; 25: 3192–203.

[38] Jin C, Yao M, Liu H, Lee CF, Ji J. Progress in the production and application of n-butanol as a biofuel. *Renewable and Sustainable Energy Reviews* 2011; 15: 4080–106.

[39] Abdullah Z, Suhaimi H, Abdullah A, Taufik MF, Marwan AG. Effect of Pentanol-Diesel Fuel Blends on Thermo-Physical Properties, Combustion Characteristics, Engine Performance, and Emissions of a Diesel Engine. *International Journal of Automotive and Mechanical Engineering 2018;* Volume 15, Issue 3 pp. 5435-5450.

[40] Huang H, Delin Lv, Zhu J, Chen Y, Zhu Z, Pan M, Huang R, Jia C. Development and validation of a new reduced diesel/n-pentanol mechanism for diesel engine applications; *Energy Fuels* 2018, 32, 9934−9948.

[41] Ramesha A, Ashoka B, Nanthagopala K, Pathyb MR, Tambarea A, Prajwal M, Phukea P, Patila S, Subbarao R. Influence of hexanol as additive with CalophyllumInophyllum biodiesel for CI engine applications; *Fuel* 249,2019, 472–485.

[42] Sivasubramanian H, Sundaresan V, Ramasubramaniam SK, Shanmugaiah SR, Nagarajan SK. *Investigation of biodiesel obtained from tomato seed as a potential fuel alternative in a CI engine*; Biofuels-Taylor and Francis, 2017.

[43] Raza M, Chen L, Ruiz R, Chu H. Influence of pentanol and dimethyl ether blending with diesel on the combustion performance and emission characteristics in a compression ignition engine under low-temperature combustion mode; *Journal of the Energy Institute, Energy Institute,* Published by Elsevier, 2019.

[44] Kattela S P, Rao S, Raju VRK. Comparative study on single-injection and pilot injection strategies for DI CI engine fuelled with the butanol/diesel blend; *International Journal of Ambient Energy*, 2018.

[45] Pouresa M, Gopala K, Sathiyagnanam AP, Kumar BR, Rana D, Saravanan S, Damodharan D. Comparative account of the effects of two high carbon alcohols (C5 & C6) on combustion, performance and emission characteristics of a DI diesel engine; *Energy sources, part A: Recovery, utilization, and environmental effects*, 2019.

[46] Saxena MR, Maurya RK, *Effect of Butanol Addition on Performance, Combustion Stability and Nano-Particle Emissions of a Conventional Diesel Engine;* SAE Technical Paper 2018-01-1795, 2018.

[47] Ahmed I, Seesy EL, Hassan H. Investigation of the effect of adding graphene oxide, graphenenanoplatelet, and multiwalled carbon nanotube additives with n-butanol-Jatropha methyl ester on a diesel engine performance; *Renewable Energy* 132,2019,558-574.

[48] Sukjit E, Liplap P, Maithomklang S, Arjharn W, *Experimental Investigation on a DI Diesel Engine Using Waste Plastic Oil Blended with Oxygenated Fuels;* SAE Technical Paper 2017-24-0116, 2017.

[49] Yanga K, Wei L, Cheung CS, Tang C, Huang Z. The effect of pentanol addition on the particulate emission characteristics of biodiesel operated diesel engine; *Fuel* 209, 2017, 132–140.

[50] Mwangi J K, Lee WJ, Tsai JH, Wu T S. Emission Reductions of Nitrogen Oxides, Particulate Matter, and Polycyclic Aromatic Hydrocarbons by Using Microalgae Biodiesel, Butanol, and Water in Diesel Engine; *Aerosol and Air Quality Research*, 15: 901–914, 2015.

[51] Yilmaz N, Atmanli A, Vigil F M. Quaternary blends of diesel, biodiesel, higher alcohols, and vegetable oil in a compression ignition engine; *Fuel* 212 2018, 462–469.

[52] Akar M A. Performance and emission characteristics of compression ignition engine operating with false flax biodiesel and butanol blends; *Advances in Mechanical Engineering* 2016, Vol. 8(2) 1–7.

[53] Chen L, Raza M, Xiao J. Combustion Analysis of an Aviation Compression Ignition Engine Burning Pentanol−Kerosene Blends under Different Injection Timings; *Energy Fuels* 2017, 31, 9429−9437.

[54] Ma Y, Huang R, Huang S, Zhang Y, Xu S, Wang Z. Experimental investigation on the effect of n-pentanol blending on spray, ignition, and combustion characteristics of waste cooking oil biodiesel; *Energy Conversion and Management* 148,2017, 440–455.

[55] Siagos I Z, Antonerias A, Karonis D. Impact of Ethanol and n-Butanol Addition on Fuel Properties and Exhaust Emissions of a Stationary Diesel Engine; *J. Energy Eng*., 2018, 144(5): 04018052.

[56] Prabakaran B, Vijayabalan P. Evaluation of the performance of n-butanol–ethanol–diesel blends in a diesel engine; *International Journal of Energy for a Clean Environment* 17(1), 81–97,2016.

[57] Dev S, Divekar P, Yanai T, Chen X. *Hydrocarbon Speciation of Diesel Ignited Ethanol and Butanol Engines*, SAE Technical Paper 2016-01-0773, 2016.

[58] Krishnamoorthi M, Malayalamurthi R. A Comparison of Performance and Emission Characteristics of Three Alcohol Biofuels: Ethanol, n-butanol, and Isopropyl; *Int. J. Pharm. Sci. Rev.Res.,* 40(2), September -October 2016; Article No. 25, Pages: 115-121.

[59] Sivasubramanian H. Performance and emission characteristics of papaya seed oil methyl ester-n-butanol-diesel blends on a stationary direct-injection CI engine, *Biofuels,* 9:4, 513-522, 2018.

[60] Zhu L, Xiao Y, Cheung CS, Guan C, Huang Z. Combustion, gaseous and particulate emission of a diesel engine fueled with n-pentanol (C5 alcohol) blended with waste cooking oil biodiesel; *Applied Thermal Engineering*, 2016, 73–79.

In: Applied Soft Computing Techniques … ISBN: 978-1-53618-180-7
Editors: Amit Kumar Thakur et al.

Chapter 6

SPECIFIC SOFT COMPUTING STRATEGIES FOR EVALUATING THE PERFORMANCE AND EMISSIONS OF A SPARK-IGNITION ENGINE USING ALCOHOL-GASOLINE BLENDED FUELS: A COMPREHENSIVE ANALYSIS

Amit Kumar Thakur[1,*], Ajay Kumar Kaviti[2], Rajesh Singh[3], Anita Gehlot[3] and Roopesh Mehra[4]

[1]School of Mechanical Engineering, Lovely Professional University, Phagwara, Punjab

[2]School of Mechanical Engineering, VNRVJIET, Hyderabad, India

[3]School of Electronics Engineering, Lovely Professional University, Phagwara, Punjab

[4]School of Mechanical Engineering, Tsinghua University, Beijing, China

ABSTRACT

The enormous consumption of fossil fuels has created an unprecedented situation, and, with the accompanying increase in the number of cars, the level of pollution has been far beyond human control. As the level of pollution has surpassed all levels, it has become essential to find an alternative fuel that can help to reduce exhaust emissions and efficiency. Experiments carried out on the SI engine are considered time-consuming, and the costs incurred in carrying out these experiments are very high. So the need for soft computing techniques is becoming extremely important in this area.

Soft computing has shown great potential in providing researchers with the exact solution that validates or predicts performance and emission parameters. The different software computing methods widely used include the Adaptive Neuro-Fuzzy Inference System (ANFIS), the Artificial Neural Network (ANN), the Fuzzy Expert System (FES), Response Surface Methodology (RSM) and Support Vector Machine (SVM). The only objective of this effort is to carry out a comprehensive review of the various researchers who have worked on soft computing techniques for SI engines with a variety of alternative fuels.

Keywords: soft computing, SI engine, ANN, ANFIS

INTRODUCTION

In the present scenario, I.C engines have generated high relevance for the comfort of human beings. Automobiles, ships, motorcycles, aircraft, locomotives are all driven by the most common power source. The I.C engines have the advantage of an enormous power-to-weight ratio. All engine varieties depend on one type or another of fuel being used [1]. With the rapid growth of the industry, fossil fuel consumption is slowly rising, and modernization plays a significant role in depletion [2].

Energy availability and energy demand will vary over the next 50 years due to the gradual decline in fossil fuels. Harmful pollutants discharged into the atmosphere have deteriorated the environment to a great extent, and great need is to reduce or maintain these emissions within safe limits. A great deal of research has been done in this respect to find the best suitable alternative fuel, which increases the performance parameters of the engine and helps in reducing the emission parameters [4-6]. It is worth noting that the experimental process for evaluating the efficiency and exhaust emissions is quite complex and involves a lot of time and money. It is indeed a complex process that can be simplified by various mathematical modeling techniques. It is effortless for the modeling technique to predict the performance characteristics of the engine, including the emissions of pollutants. The added advantage of their high accuracy and less time-consuming soft computing has become an eye-catcher in this field. Optimization of all soft computing techniques requires minimum effort. Thus it has gained a great deal of respect from various researchers who are now using this technique to generate data quickly and efficiently. Soft computing is a problem-solving technique that uses an approximate model to provide a better solution to complex problems. Soft computing has made its way in this field with the sole intention of imitating a copy of the human brain. The modeling technique has proved to be beneficial in accurately diagnosing engine malfunction, ignition activation, heat release rate, and significant progress in the control system. Indeed, it has shown the path to better learning in a more unwavering way to gain an upper edge over the experimental process, making way for research shortly. Various soft computing or modeling techniques are of extreme importance, and more work is being done on these modeling techniques so that better results are delivered. Different soft computing techniques used to predict and optimize engine parameters and release harmful gases [7]. This study aims to carry out a comprehensive review of the various researchers who have worked on soft computing techniques for SI engines with different alternative fuels.

VARIOUS SOFT COMPUTING OR MODELING TECHNIQUES

Methodology of Anfis

A Fuzzy Control Algorithm is an approach based on truth or rather than a standard true or false approach. It is a method of interpolation of different vector variables. The Fuzzy Control Algorithm can solve any problem. Functions already effectively performed by humans can be more easily automated by the use of the Fuzzy control algorithm. Using Fuzzy logic capricious and complex non-linear functions are modeled, and it provides an alternative to nonlinear modeling. The inference process, the membership function, and the various rules of Fuzzy logic facilitate the handling of non-linearity and complex problems, resulting in improved performance, reduced design costs, and more straightforward implementation. Neuro-fuzzy makes the most of the learning features of the neural network to process that part of the fuzzy data independently. Neuro-fuzzy consists of fuzzy nodes rather than plain input and output nodes and is called an associative memory model. Training in a separate network is better than the entire network. Jagnis was a genius behind the Neuro-Fuzzy Inference (ANFIS) process. Fuzzy systems are a quick learner of parameters. Using a hybrid learning algorithm, ANFIS can use human knowledge and numeric data to map input-output. The fuzzy modeling method uses a neuro-fuzzy technique that obtains data set information so that membership (MF) parameters can be determined. The respective input and output data are, therefore, controlled using a fuzzy inference system. ANFIS has a similar configuration to the traditional neural network. Figure 1 shows the basic flow diagram.

There are five levels of ANFIS architecture. The ANFIS data processing approach is a two-input system. The system has only one output as z with two inputs like x and y. Therefore, if the underlying rule consists of two fuzzy rules, it may be represented as Eqs. (1) and (2), respectively.

$$\text{If } x \text{ is } A1 \text{ and } y \text{ is } B1, \text{ then } f\,1 = p1x + q1y + r1 \tag{1}$$

If x is A2 and y is B2, then f2 = p2x + q2y + r2 (2)

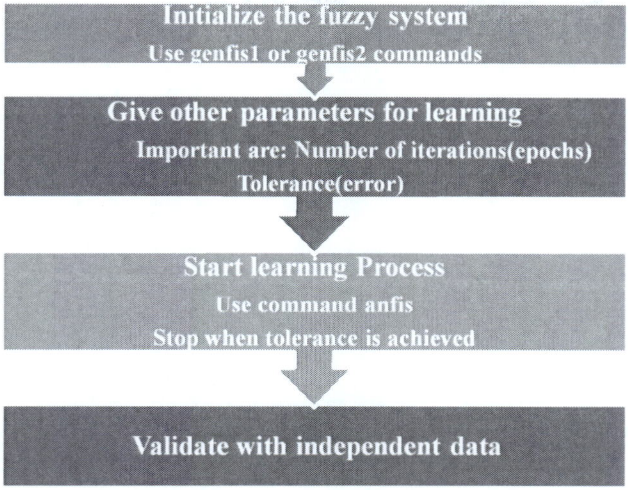

Figure 1. The basic flow diagram in ANFIS.

For each input in this system, there are two fuzzy sets like A1, A2, and B1, B2. The circle represents the fixed nodes, as shown in Figure2, while the adaptive nodes are square. The design parameters allocated to ANFIS for training were p, q, and r with f as the output.

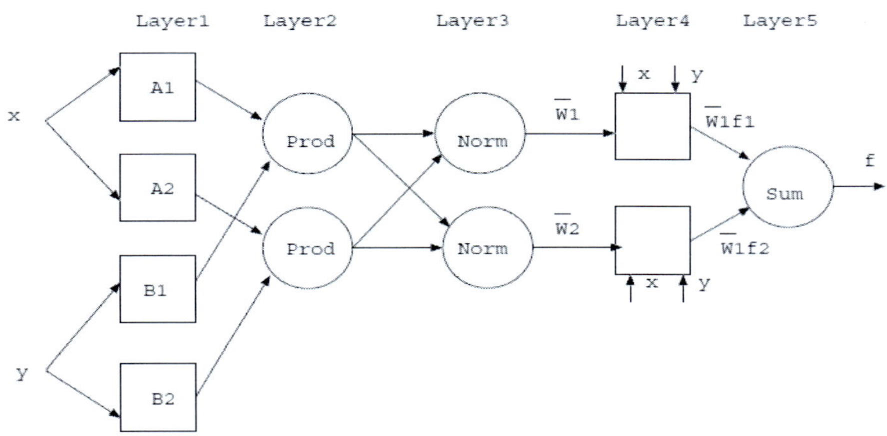

Figure 2. Equivalent ANFIS structure [11].

Layer 1: In this layer, every Node, denoted by *i* is a square node with the output function of

$$O_i^1 = \mu_{Ai}(x) \text{ and } O_i^1 = \mu_{Bi}(y) \tag{3}$$

The membership functions $\mu_{Ai}(x)$ $\mu_{Bi}(y)$ determine the degree to which the specified inputs x and y comply with A and B of the linguistic labels. 0 and 1 are the values selected by $\mu_{Ai}(x)$ and $\mu_{Bi}(y)$.

Eqs. (4) and (5) are respectively used for generalized bell-shape and Gaussian functions.

$$\mu_{Ai,Bi}(x) = 1/\left(1 + \left[((x - c_i)/a_i)^2\right]^{bi}\right) \tag{4}$$

$$\mu_{Ai,Bi}(x) = \exp\left[-((x - c_i)/a_i)^2\right] \tag{5}$$

$\{a_i, b_i, c_i\}$ is the parameter set.

Layer 2: This layer determines the strength of the associated rule, which takes in the fixed circular nodes.

$$O_i^2 = \mu_{Ai}(x) \times \mu_{Ai}(y) \; i = 1, 2. \tag{6}$$

Layer 3: This N-labeled circular node determines the ratio of fire-power of the rule to the sum of the fire-power of every rule:

$$O_i^3 = w_i/(w_1 + w_2), \; i = 1,2. \tag{7}$$

Layer 4: Node function is in the square form as in Eq.8.

$$O_i^4 = \overline{w}_i f_i = w_i(p_i x + q_i + r_i) \tag{8}$$

w_i is the output of layer 3 and (p_i, q_i, r_i) is the parameter set.

Layer 5: This layer is a circular node, and the sum of all input signals is the overall output.

$$O_i^5 = \sum \overline{w}_i f_i = \sum_i w_i f_i / \sum_i w_i \tag{9}$$

Thus there is an adaptive five-layer network [11]. In the fuzzy control toolbox, ANFIS finds the parameters in the fuzzy system that the optimization scheme best fits the data. This neural network uses any optimization methods to calculate the gradient for almost all of the optimization algorithms. Due to the unclear picture of what the initial fuzzy system should be in terms of the number of membership functions, the GENFIS1 command initiates, as shown in Figure 3. This command finds a sound starting system and analyzes the data in a very crude way. So, GENFIS 1 should be used to generate the first FIS system. We can then include ANFIS to create the best FIS system. To begin the optimization process, ANFIS is used, which uses the result generated by GENFIS 1. The display of the ANFIS Editor divides itself into four main sub displays: 1. Loading the data, 2. General of the FIS, 3. FIS Train, 4. Testing FIS.

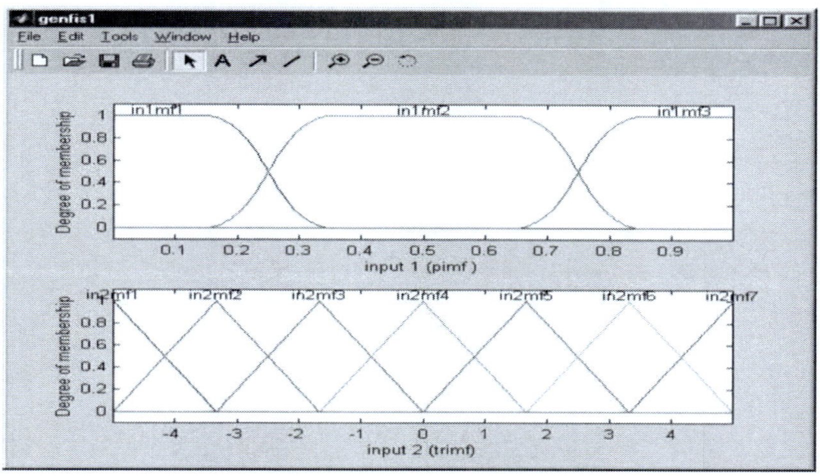

Figure 3. GENFIS1 toolbar.

Load Data

This section provides a selection of the type of data (training, testing, checking, or demo). Training data comes as by default, which becomes the first choice. Data gets generated and displayed in a matrix form on the MATLAB order. The first column consists of data from the input, and the last one consists of data from output. Locating the data is an essential aspect and needs proper attention. The disk or the workspace stores the data. As shown in Figure4, it demands the input variable name in the given space once the load data set gets clicked.

General FIS

The first fuzzy structure is generated by the command Generate FIS when the data is loaded. The Grid partition remains on the editor as the default choice. The following window will open as the Generate FIS button is clicked. The Gbellmf membership functions are selected by default, as shown in Figure 5.

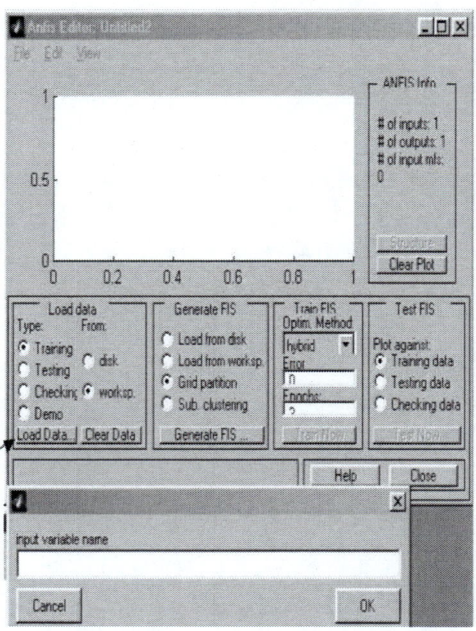

Figure 4. ANFIS editor.

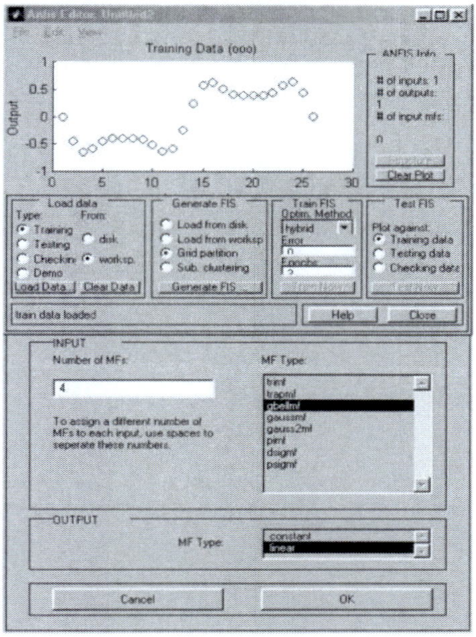

Figure 5. Choosing a membership function.

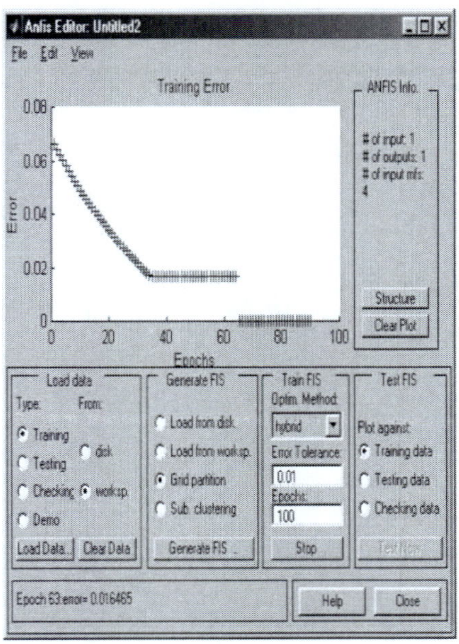

Figure 6. Training process window.

Train FIS

The training block is the key computing field. The hybrid algorithm is used, with a tolerance error of 0. In the case of incomplete fit, the error tolerance changes to 0.01. Training error occurs during the training process. The epochs are the number of iterations remains at 100. The training process initiates with the train button, as shown in Figure6.

Test FIS

Once the results are within limits, testing of FIS performance follows. The default value is the training data, and on clicking the test now button, the following displays on the screen, as in Figure 7.

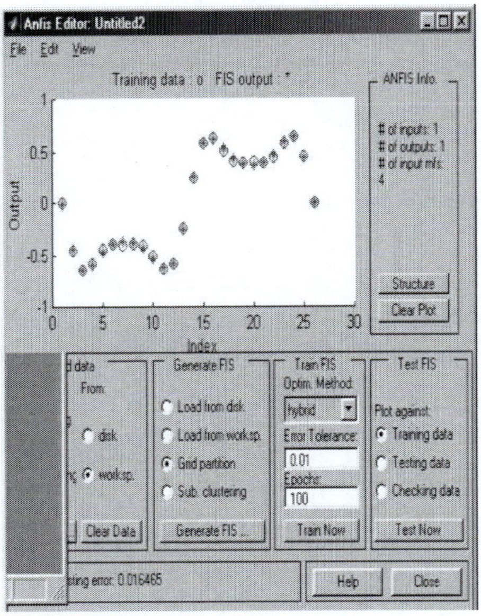

Figure 7. Testing process window.

Review of Previous Work

Work on different ethanol-gasoline blends was carried out by Najafi et al.[2] using SVM and ANFIS to analyze the performance and emission parameters of the SI engine. The test was performed at different speeds and under 45 different test conditions. Brake Specific Fuel Consumption (BSFC) decreased as the percentage of ethanol increased. Further improvement in BP, engine torque, BTE, and volumetric efficiency was observed compared to gasoline oil. With ethanol mixtures, the concentrations of carbon monoxide and hydrocarbon decreases, but the levels of CO_2 and NOx increases. Test data predicted emissions and engine performance. The observations for the optimum value is appropriate for SVM modeling based on the specific kernel width and penalty parameters (C) of the RBF (Radial Base Function). Gaussian curve membership (Gaussmf) and 200 epochs are the best choices for the ANFIS modeling training process. The correlation coefficient(R) and accuracy for SVM and ANFIS decreased by 0.660–1, 65.310–99.330 percent, and 79.270–98.810, respectively. SVM and ANFIS are capable enough to predict engine performance and emissions, although ANFIS won the upper hand. On the other hand, SVM dominated ANFIS in terms of computing. The difference between the performance of ANFIS and SVM can be subject to different theoretical backgrounds. SVM and ANFIS scored performance over ANN when a comparison was made in this case study by the researcher.

Research indicates that ANFIS is used to detect performance and emission parameters using isopropanol macro-emulsions of petroleum-water. The predictive ANFIS algorithm with active pressure braking is the method of learning and helps in supplying the experimental data. This research aimed to offer a robust model that could demonstrate different performance characteristics by using different gasoline liquid macro-emulsions. Providing quick data and results, an acceptable ANFIS model was modeled and used in experimental design procedures. The exhalation value generates on the speed of the engine at full load. Petrol-water macro emulsions, including iso-propanol, drive the motor. Current and forecast exhaust emissions display a good relationship between the actual and the projected values. As a result, ANFIS has proved to be a valuable tool for predicting exhaust emissions for SI engines with a knack to solve complex problems and save time directly [11].

ANN Methodology

Much work on ANN's comprehensive theory, principles, and working philosophy is published, which gives a detailed description of ANN. However, this section provides a simple introduction of ANN and its benefits. ANN is a common technique in machine learning in different fields. This black-box modeling method provides the said relationship between inputs and corresponding outputs. The correct input and output are selected without loss of accuracy so that predictions should be relevant and accurate. ANN boasts a higher level of competence in the approximation of the input-output association due to its ability to capture the non-linear behavior of a given system, particularly for a large number of calculated data. Using the perceptron method, Rosenblatt developed the principle that the activation function forms the basis of the output model, which defines the threshold value and the weights associated with the input. Input(s), additional frame, Activation function, and output were ANN components. The weight is a value that specifies the input intensity of the node. A bias function manages the activation function.

$$u = b + \sum_{i=1}^{N} w_i x_i \tag{10}$$

Activation function checks and validates the performance of the perceptron model. ANN uses the transfer function to design the tansig function of the ANN, which comes from the logsig function. This study shows that tansig learns much faster than logsig. The Tansig function yields exceptional analytical advantages because it assumes the value of logsig between-1 and 1 rather than 0–1. Researchers investigated the efficiency and emission characteristics of different fuels using ANN in their study. Correctly, the multi-layer perceptron (MLP) network provides ample support. MLP performs the computation with one or more hidden layers of neurons. The weighted power of the first hidden layer comes as a reference for either the output neurons or the additional hidden layers.

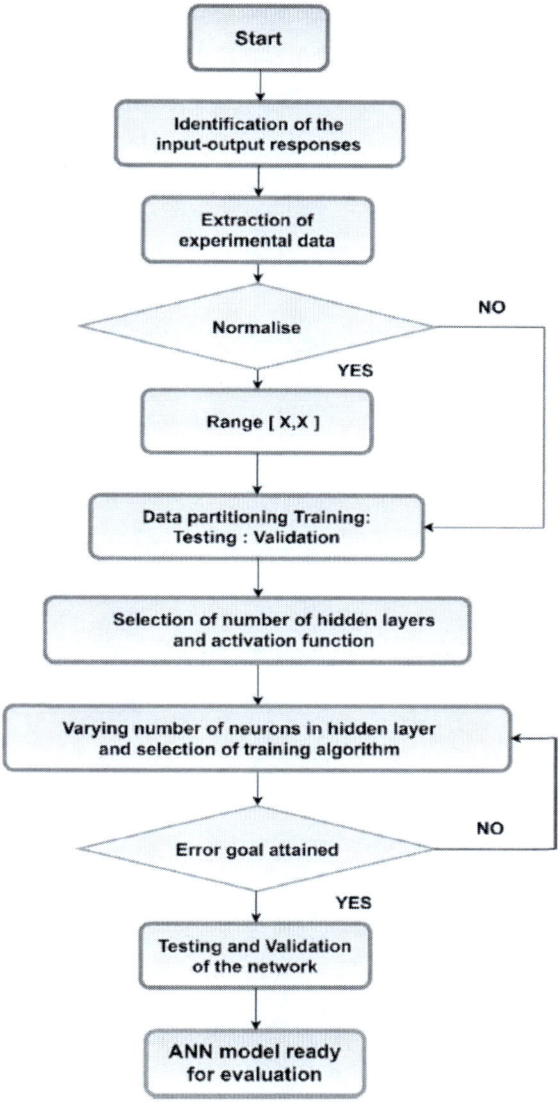

Figure 8. ANN implementation flow chart [8].

The growth of hidden layers improved the ability of the network to monitor non-linear behavior, especially when a large number of layers were present. The basic structure of the ANN topology consists of four input nodes, five hidden layer neurons, and two output neurons. The three main phases of the ANN are the selection, learning, and evaluation of parameters. The input-output data is processed repeatedly by standardization within a given range. The information is then randomly partitioned in terms of training, assessment, and verification. Choosing input and output variables, the number of hidden strata, the number of neurons in the cached layer, the activation function, and the training algorithms as the primary parameters characterize the ANN test. Learning algorithms are responsible for changing weights and biases so that predictions can be made persuasively in the light of reasonable error limitations. Once the learning process completes, the evaluation and validation process becomes trivial as it provides a valuable approach to the efficiency of the ANN models. The specific preparation, evaluation, and analysis of the coefficient of correlation, R, and the coefficient of decision, R2 shall be investigated. A new set of input data is added to the network to validate it. The additional statistical analysis assesses the effectiveness of the expected responses. The mathematical method used is mean relative error, root mean square error, mean absolute percentage, and mean percentage error. Figure 8 and 9 demonstrate the implementation steps of the ANN following the ANN topology [8].

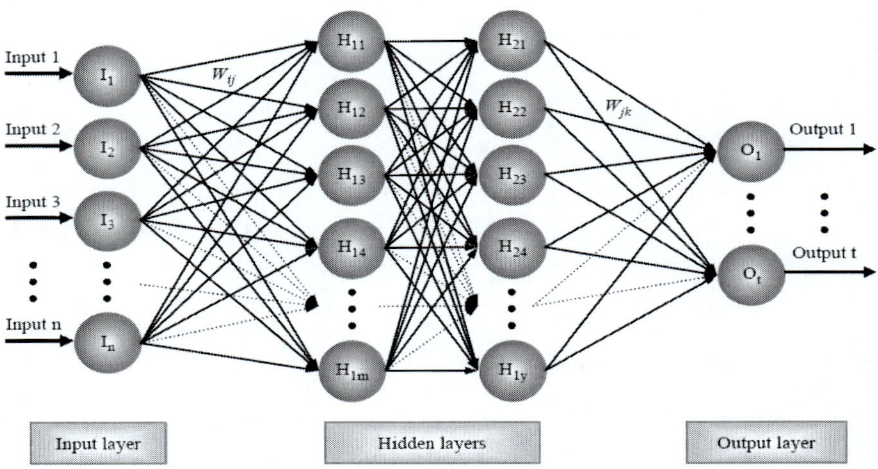

Figure 9. ANN topology [8].

Review of Previous Work

In this research study, ANN is used to predict power plant, torque output, and exhaust emissions from a4-cylinder, 4-stroke spark ignition (SI) engine. Ethanol was mixed in different fuel proportions (0, 5, 10, 15, and 20 percent) and used to collect training and testing data. ANN quality verification carries out in conjunction with the predictive data for experimental results using a standard backpropagation algorithm with few experimental training data. This study found that ANN had the best reliability in predicting emission and efficiency parameters with a correlation coefficient of 0.71, 0.90, 0.96 and 0.98 for NOx, HC, CO_2 and CO, and 0.96 and 0.99 for brake power and torque respectively.

ANN wins over numerical and mathematical methods as the prediction is very fast, accurate, and reliable. It is a powerful modeling technique with a high edge for predicting output and emissions. The results of this investigation showed a fine mapping of the problem entrances and outputs. The regression coefficients are higher when both the predicted and measured regression lines were formed in the data set [1]. The ANN and FES predict the power of the engine, torque, amount of fuel used, and HC emission in the gasoline engine. Data collected from the researcher's experimental work gives accurate predictions. The opening of the intake valve and the speed of the motor are the input parameters. The output parameters to be determined are power output, torque, brake specific fuel consumption, and hydrocarbon discharge. Comparison of the T-test and regression analysis in MATLAB with the investigation statistics and the results for ANN and FES is the main aim of this experiment. For the confidence interval of $p>0.05$, both groups of data did not differ, and the differences are not statistically significant. The result showed that ANN and FES are beneficial for the automotive industry and technology instead of experimental work. Compared to conventional methods, ANN and FES have the most significant advantage of being quick in calculations, simple in approach, and thus requiring no experimental work to reduce engineering efforts [5]. An artificial neural network (ANN) model predicts the quality and emission parameters for an n-butanol-gasoline-powered SI engine. Relative to the experimental data with the expected results, the mean relative failure ranged from 0.1943 to 9.9528%, and the correlation coefficient ranged from 0.9929 to 0.9996. The excellent precision of the artificial neural network helped predict the combustion of n-butanol-gasoline mixtures. The addition of butanol to gasoline under stoichiometric conditions has led to a reduction in BTE. B30 showed excellent values based on similar BTE and lower CO emissions (8.7%), UHC emissions (15.3%), and NOx emissions (6.4%) than G100 emissions. Once again, ANN demonstrates its superiority as R is very close to 1. The RSME and MRE values are found to be very small, thus proving to be the best tool for comparison and simulation of engine parameters [10]. The focus of this study was on predicting engine power, torque, and exhaust emission with gasoline and ethanol blends in varying proportions using ANN (artificial neural network). Model training carries out using the standard back-propagation algorithm. A comparison of test results provides the accuracy of the ANN simulations. ANN modeling verifies and predicts the outcomes of output parameters such as engine torque, exhaust emissions, exhaust gas temperature, and BSFC keeping the input parameters as fuel type and engine speed. ANN model effectively forecast engine performance, exhaust emission, exhaust gas temperature, and BSFC with the approval value of the correlation coefficient. The best results in parameter prediction are a 2–5–5–7 network algorithm learning architecture. Regarding training and test data, the R2 value after training is very close to 1, and the RMS value was less than 0.0155 for the test data, which is within acceptable limits [14]. The findings of this study are the efficiency of the SI engine operating from pure gasoline to pure ethanol using the artificial neural network (ANN). The investigation reveals low BSFCs with amplified BTE and volumetric efficiency. Emissions of CO and HC were reported to be on the smaller side but increased compared to gasoline as a single fuel of CO_2 and NOx. The input parameter introduced in the ANN model was the non-loaded engine load, 25, 50, 75 percent, and full load with mixed fuel at an engine speed of 2500 rpm. Validation carries out at 15 percent of the experimental data. Seventy percent of data is for training, and the last 15 percent is to see an increase in the results of generalization of the network.

The correlation coefficient ranged for all output variables, and exhaust emissions showed better values. RSME is very small, and the value of MRE in the domain decreased to 0.12-5.56 percent. R values did not rise when there were more than 20 neurons selected [17]. The4-cylinder SI KIA, 1.3 SOHC mixture of 0-20 percent ethanol and an increase of 5 percent, was performed by Najafi et al.[19]. The ANN model evaluates the emission and efficiency parameters of the pollutant. The input was the engine speed that varied with an interval of 1000rpm from 1000rpm to 5000rpm. The efficiency of the system included emissions from exhaust engines and engine performance parameters. Coefficient of correlation, R evaluates the tansig, and logsig activation functions. Training carries out at 70 percent of experimental data with verification of 30 percent of experimental data. The performance of the RSME and MRE prediction models are determined to be properly assessed. The best forecast for peak R of 0.9999 was the layout of 20-secret ANN neurons trained under the trainlm training system. R was found to be 0.999, 0.995, 0.981, 0.985, 0.986, 0.994, 0.985, 0.985, and 0.983 for BP, engine torque, BTE, volumetric efficiency, BSFC, carbon monoxide, CO_2, hydrocarbon and NOx. The MRE value of the forecast evaluated ranged from 0.46 percent to 5.57 percent, and the RSME was lower. Cay et al. [20] using petrol and methanol at various torques and engine speeds to carried out his research findings. ANN model forecasts BSFC, CO, HC, and AFR of the SI engine. To determine the finest quantity of neurons producing the lowest cost function value, i.e., MSE neurons varied from 5 to 15. Work also included four separate models with four different learning algorithms to predict the above values. The input and output data are standardized for analysis between 0–1.15 percent of the total investigation data set, and the remaining 85 percent makes its way for training purposes. Results revealed the best forecast for carbon monoxide, hydrocarbon, BSFC, and air-fuel ratios. The R2 for CO, HC, BSFC and AFR training and testing is 0.9969, 0.9993, 0.9996, 0.9998, 0.9776, 0.9984, 0.9986 and 0.9961. Kiani et al. [21] executed a backpropagation ANN model in his research work to foretell engine efficiency and pollutant production from the SI engine fueled by ethanol-gasoline fuel mixtures. Torque, power, CO_2, HC, and

NOx are the expected output parameters, with input parameters varying from 0% to 20% and a 5% motor speed interval and varying load conditions (25%, 50%, 75%, and 100%). Inputs include several ethanol-petrol blends. Various training algorithms give a better value of MSE and R, i.e., trainlm, traingdx, trainscg. To make out the optimum neuron number, the number of neurons per layer often ranged from 10 to 30. The model test carries out at 20% of the total experimental data, and the remaining 80% comes in handy for training purposes. Applied with tansig activation function, 0.9998 was the value of R obtained, and it was trained in the 2-layer architecture of 3-25-25-6 using trainscg algorithms. The tests for torque, brake, CO, CO_2, Hydrocarbon, and NOx in the ANN model correlated with 0.99, 0.96, 0.98, 0.96, 0.90, and 0.71, respectively. Kapusuz et al. [22] used a variable compression ratio of TD-43, petrol engine with a mixture of methanol and ethanol with unleaded gasoline to determine the performance parameters. ANN investigated the output parameters of the engine, namely torque, brake strength, and BSFC. Fuel blend occurs in the volumetric ratio of 5, 10, and 15% with varying engine speed. The efficiency parameter of the engine becomes the output with the speed of the motor and the fuel blend ratio as an output parameter. Based on the output parameters, three ANN models give an R-value between 0 and 1. The training with 85% of the total investigational data, and 15% data for the testing process gives accurate results. R-value near to 1 suggests full compatibility between the model of prediction. The network architecture shows the best torque, brake power, and BSFC prediction results using the tansig transfer function.

Methodology of SVM

Vapnik has developed SVM as a supervised learning system based on quantitative theory. SVM is now widely used with its ability to solve learning problems for a variety of purposes, such as detection, pattern recognition, and classification. SVM has excellent potential for widespread use as it incorporates the theory of structural risk minimization, which reduces the higher boundary of error in generalization. Assuming a set of training data $\{x_i, t_i\}$, i = 1, 2,..., n, $x_i \in R^d$, $t_i \in R$, where $x_i$is the input data, and t_i is the corresponding target value. The SVM uses the following linear equation to determine the target values:

$$y_i = w^T x_i + b$$

If yi is the SVM output, w is a vector of d dimensions, and b is a scalar. By solving the following optimization problem using an ε - insensitive loss function, the optimal linear regression function results.

$$\min \frac{1}{2} w^T w + C \sum_{i=1}^{n} (\xi_i^+ + \xi_i^-)$$

$$\text{subject to} \begin{cases} -\varepsilon - \xi_i^- \leq t_i - y_i \leq \varepsilon + \xi_i^+ \\ \xi_i^+ \geq 0 \\ \xi_i^- \geq 0 \end{cases} \quad \forall i$$

C is a penalty parameter, ε is a pre-specified value, and ξ_i^+,$\xi_i^-$are slack variables representing the upper and the lower constraints on the system's output. The SVM maps data in a high-dimensional nonlinear regression function field. It produces linear returns by a kernel function, which states *that,* $K(x_i, x_j) = \phi(x_i)\,\phi(x_j)$, where ϕ is a nonlinear function. The formula below evaluates SVM.

$$y_i = \sum_{i=1}^{n} (\alpha_i^+ - \alpha_i^-) K(x_i, x) + b$$

α_i^+, α_i^-,are Lagrange multipliers [2].

Review of Previous Work

SI engine fuelled by butanol gasoline blends analyzes the performance and emission values under various equivalents. B30 displayed outstanding engine performance results and emission parameters with CO (2.3 percent - 8.7 percent), UHC (12.4 percent - 27.5 percent), and NOx (2.8 percent - 19.6 percent) compared to gasoline. Butanol has shown excellent results as a strong contender for petrol with the ability to reduce contaminants. Since the engine test process is slow and costly, the support vector regression (SVR), predicted accurate results. R and MRE values are between 0.9940 and 0.9998 and 0.1901– 10.2570 percent, respectively. The prediction of BTE, BSFC, carbon monoxide, unburned hydrocarbons, and NOx has been substantial [12]. This work suggests the use of LS-Support Vector Machine to create a set of SI engine torque models using sample experiment data and different engine speeds.

Nonetheless, a re-training system can take time. Proposal for using an exceptional algorithm for improving the LS-SVM method forms the base of the experiments. With new training data, it is possible to successfully adjust torque models built of the incremental LS-SVM to allow the model to develop continuously. The study revealed that the forecasted values are consistent with the actual test results, with the new LS-SVM torque versions. The experimental results also indicate that the exactness of the LS-SVM incremental engine performance and scratch retraining performance models is almost similar. However, the former has a much shorter time than the latter one. In the retraining process, the overall training time saved through an LS-SVM is crucial [16].

Methodology of RSM

RSM (response surface methodology) handles the relationship linking numerous illuminating variables and one or more variables very effectively and efficiently. The inspiration behind RSM is to acquire the most favorable rejoinder using the correct string of the expected experiment. Box and Wilson have proposed to achieve this 2nd-degree polynomial method. RSM (response surface methodology) combines the scientific and statistical purposes to study the functional relationship with the response

(y) and the input variable(x1,x2,---xk). This relationship is uncertain, approximates itself by a small model of polynomial form expressed as:

$$Y = f'(x)\beta + \varepsilon$$

β refers to parameters, and ε is a random experimental error. Quantity f(x) β symbolizes the mean response. RSM uses two essential models frequently.

$$y = \beta_0 + \sum_{i=1}^{k} \beta_i x_i + \varepsilon$$

And the second-degree model (d =2):

$$y = \beta_0 + \sum_{i=1}^{k} \beta_i x_i + \sum\sum_{i<j} \beta_{ij} x_i x_j + \sum_{i=1}^{k} \beta_{ii} x_i^2 + \varepsilon$$

Considering a model like (1) is intended to be as follows:

1. An approximate relation forms between y and x1, x2, ..., xk, to know the response values for the control variables set.
2. To find out the importance of factors identified by x1, x2, ... xk by hypothesis checking.
3. To evaluate the optimum settings of x1, x2, ... xk, which gives the highest or lowest value over a particular area of interest[3].

Review of Previous Work

RSM (response surface methodology) is used in this prediction to forecast engine performance and emissions of SI (spark ignition) from ethanol and gasoline blends. Brake power, engine torque escalated while BSFC decreased. The motor was powered at different speeds to meet 45 different conditions. Carbon monoxide and hydrocarbon emanation have declined with the advent of ethanol mixtures, but CO_2 and NOx emissions have risen to a higher value. Using the desirability approach of RSM, the definitive objective was to trim down emissions and maximize the performance optimization of independent variables. In scheming the experiments, one of the most important statistical tool DoE (experiment design) based on RSM comes in picture. The RSM desirability approach optimizes engine operating parameters. It was noted, after careful examination, that the performance limit of ethanol-petrol blends was similar to gasoline, but the exhaust emissions increased considerably. When the engine was running at 3000 rpm, E10 established itself as the best candidate. The results from the complete investigation reveal that the value of BSFC, torque, brake power is 0.25(kg/kW hr), 103.66 (Nm), 35.26 (kW).

The value of NOx, HC, CO_2, CO is 1300 (ppm), 136.6 (ppm), 12.8 (percent Vol.), and 3.5 (percent Vol.) respectively at optimum input parameters. In improving engine performance and exhaust emissions, RSM thus proved to be a strong contender. The design of RSM-based experiments is exceedingly fruitful in scheming the research. Statistical investigation recognizes the most critical constraint affecting the parameters of emission and performance. When the engine was running at 3000rpm, E10 produced fewer emissions. This analysis gives a good value of desirability using the RSM method in terms of the simplest and most effective strategy. E10 had high appropriateness of 74% at 3000rpm, resulting in the most optimal conditions. The significant contribution of this research provides a pathway for other researchers to develop a new approach. It also provides users with added support to predict engine efficiency and to determine emissions.

To sum up, this study exposed a wide range of engine speeds and loads using ethanol-gasoline blends. With less complicated experimentation, the RSM methodology provided greater flexibility and accuracy. Besides that, the surface of the response and the mathematical models will predict the results in detail. RSM methodology leads to the fact that fewer experiments result in lower costs for all future research and consumes less time. The strength of the RSM is that it extends to any form of the engine [3].

The automotive industry's biggest challenge today is to find the right optimum steadiness linking engine feat and exhaust emanations. This study explores the correct use of RSM to analyze the engine performance and exhaust secretion from an SI engine by using gasoline- butanol blends in the volumetric ratio of 5, 10, and 15%. The experiment was conducted at different engine speeds, resulting in 13 different conditions. DoE (experiment design) has again been used to optimize the independent variables. With the sole aim of helping to achieve the performance parameters and reduce emissions, the desirability approach finds its place in performing experiments. With an increase in BSFC along with an increment of 15 percent for 2-butanol in the blends gives higher values of brake power, BMEP and BTE.

As shown by the RSM model, the 2-butanol charisma of up to 15 percent has helped to achieve lower emissions of nitrogen oxides, carbon monoxides, and unburnt hydrocarbons. In contrast, carbon dioxide emissions have risen to a high level. GBu15 blend is the optimum blend to generate the best output and emission characteristics compared to other combinations using the desirability method. RSM's desirability approach proves to be the best technique for optimization. The desirability for optimum engine performance and exhaust emission increases by 0.8. The engine performance for GBu15 results in lower brake power, BMEP, and BSFC. GBu15 provided lesser NOx, CO, and HC but elevated CO_2 by 17 percent, 26.9 percent, 25.7 percent, and 5.9 percent, respectively, as this research work took. The models developed using RSM are adequate, and the prediction error for brake power, BMEP, BSFC, BTE, NOx, and CO_2 are less than 5 percent.

Nevertheless, the error rose to 10 percent, and CO and HC emissions are a touch higher. Overall, RSM optimization methodology is useful for any fuel mix. It helps to find the best blend ratio to achieve the best engine efficiency and low exhaust emissions. Figure 10 depicts the process of engine optimization using RSM. Then experimental design proved to be trustworthy in reducing the time by curtailing the number of experiments to be carried out and thus reducing the cost of these experiments [7].

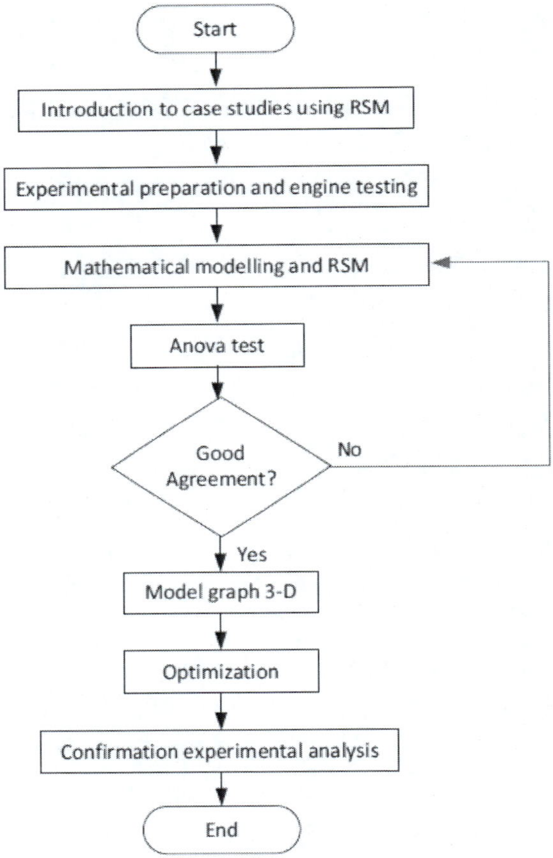

Figure 10. Flow chart of engine optimization using RSM [7].

This thesis discusses the work carried out in the field of SI engine performance and emanation with the Surface Response (RSM) methodology in investigating the optimum ratio of fusel oil and gasoline before and after the extraction of water. Multi-target improvements carry out to optimize brake strength, BTE and minimize BSFC, NOx, HC, and CO emission. The rotating extractor method digs up the fusel oil's water content. The experimental study is carried out at various engine speeds and various throttle valves opening position (15 percent, 30%, 45%, and 60 percent). At a confidence level of 95 percent, models built for responses are statistically significant. After the withdrawal of water from FAWE20 (80% gasoline fuel, 20% fusel oil after water extraction), significant improvement occurs in the heating value, thus concluding that FAWE20 is the optimal blended fuel. Approximately 55 (percentage of WOT) for load and 4500 rpm were the best state of engine parameters with FAWE20. The study found that BTE, BSFC, NOx, HC, and CO release showed typical values with high desirability of 0.707. Depending on RSM, the DoE simply defines the parameters that have a significant impact on engine performance and emissions to support and model the experimental and statistical research. The innovative design and review by ANOVA agreed on a mutual basis that the water content derived from fusel oil improved efficiency and emissions for the SI engines. This research has opened the field to further investigation into the use of higher compression ratio fusel oil blends [9]. The author examined the impact of fusel oil- gasoline mixes on motor performance and emissions up to 20% of fusel oil. Using the ANOVA study, all models built for that reason receives a better fit. Two fusel oil and gasoline mixes, along with regular gasoline as baseline fuel, were used for this experiment. The sugar industry provided fusel oil. The ideal blend ratio is at an engine load of 60 (% of WOT) at 4500 rpm among the three studied oil. The load and speed of the engine guide the engine output and emissions.

At the same time, the combined fuels (F10 and F20) helped to reduce NOx. With the rise in the percentage of fusel oil, a decrease in engine performance occurs, suggesting that the brake power decreased, and the BSFC increased to overall value. The pure gasoline was quite insignificant in comparison with that. Brake power, BSFC, CO, and NOx gives the best desirable value of 0.852.

As predicted by the DoE and ANNOVA, fusel oil has proven to be the desired alternative fuel for SI engines. Further investigation into the effect of injection pressure, injection time, and ignition timing will open questions for other researchers [15]. Awad et al. [23] examined an SI engine that used a mixture of fusel oil and petrol. The experiment designs itself on 15 percent varying full throttle opening position, 30 percent, 45 percent, and 60 percent with 10 percent and 20 percent fusel oil in gasoline fuel and at various speeds. Results give the best possible situation to be calculated by RSM (response surface methodology). This situation arises with 20% fusel oil in gasoline fuel with more than 50% throttle opening at 4500 rpm. The BSFC, BTE CO, NOx, and HC had a significant influence on the water content derived from fusel oil. On the other hand, the fuel did not have any considerable impact on the brake power. Chen et al. [25] investigated the most acceptable ethanol blending ratio for achieving high fuel conversion and low air emissions for the SI engine. The gasoline-ethanol mixing ratio, engine speed, and throttle location are the significant factors influencing conversion efficiency and emissions from air pollution. Thus the need to optimize these factors becomes the need of the hour to reduce exhaust emissions from the engine and to ensure a rise in performance parameters. The best gasoline-ethanol mixture ratio is 3.92 to 4.12 of ethanol to gasoline by volume. This mixture is, therefore, comparable to fuel efficiency and reduces exhaust emissions, as expected with Box-Behnken's model. Yusri et al. [25-26] carried out his work on the SI engine, taking on a series of experiments involving a mixture of secondary butyl alcohol and gasoline fuel with various blends of alcohol mixture. RSM's core composite model forms the conceptual methodology. The engine speed and the secondary butyl alcohol volume ratio were the component of this study. At every data point, this core composite model analyzes 13 experimental points with three replication periods. Results revealed the best possible design of models of engine performance and exhaust emission parameters in terms of brake power, BMEP, BSFC, BTE, NOx, CO, CO_2, and HC. Secondary butyl alcohol-gasoline blends successfully boost engine performance and reduce emissions. Using the desirability approach and comparison carried out under optimal conditions displayed a higher value for brake power. RSM's core composite model became the critical feature of Najafi et al.[27] undertaking a comprehensive investigation. Research work carries out at various engine speeds from 2000-4000rpm and different blends in the varying ratios of 5%, 10%, 15%, and 20% of ethanol in gasoline. The chief intention of their inquiry was to determine the optimal situation for the superior performance and the lowest release of harmful gases according to RSM's preliminary analysis. Close investigation revealed that E90 gave the best results.

REVIEW OF PREVIOUS WORK OF FUZZY EXPERT SYSTEM (FES)

A fuzzy modeling system effectively does human reliability analysis. This analysis weighs up the prospect of flawed action. The fugacious logic model is used to produce inferences based on a knowledge base of modeling by quantifying imprecise information (fuzzy input sets). Lofti Zadeh anticipated the idea of fuzzy logic. Unlike human thinking, in terms of the degree of truth, fuzzy logic carries out its process of handling data. Handling data is done in the same way as human thinks, but fuzzy logic makes things really lightning quick. The data dealt with by fuzzy logic, which permits partial membership, are known as fuzzy sets [28]. Fuzzy logic effectively and efficiently handles unfocussed, erroneous, and indistinct data.

Based on expert knowledge, Fuzzy logic converts a linguistic modeling stratagem into a system of fuzzy logic modeling strategy[29].In their research, Kilagiz et al. [30] modeled a Fuzzy expert system which, in the very beginning, calculated the fuel consumption with the assistance of all pollutant emissions from the SI engine. Considering all due support from the estimated fuel consumption and emission values, FES concludes for any kind of fault in the fuel system. FES also identifies any flaw in the ignition system, any discrepancy in the engine, and checks whether intake and exhaust valves are in proper order. Once FES detects the defects, it recommends refurbishing the fault. Emissions are within acceptable limits at high ratios, and opinion is shared by the system to repair the flaw observed.

The deadline for the degree of membership starts at 0 and ends at 1. The members of the fuzzy set lend hands to other fuzzy group members.

The Takagi-Sugeno-Kang (TSK) and the Mamdani-like are two essential ways of using fuzzy rules successfully. Fuzzy logic is composed of four parameters, namely Fuzzification, inference, knowledge base, and defuzzification. Tasdemir et al. have taken advantage of ANN and FES modeling in their work by foreshadowing the performance parameters and pollutant emission in the gasoline engine.

The experiment aims to predict the values of engine torque, engine power, hydrocarbon emission, and specific fuel consumption. Experiments generate the data for the ANN and FES model, and these data validate the performance parameters and emissions.

The input of the model is engine speed and the intake valve opening. On comparing the test data and the prediction data from the model, the forecast values are very close to the actual values. The models employed can be used to predict engine performance and emanation with all the money needed to save time and capital investment. The fuzzy neural network blends the fuzzy logic with the neural network, which can handle massive data. It has the best of its ability to handle logical, numerical, and linguistic information. It can handle inaccurate and vague data very efficiently. It can learn, plan, and visualize itself. The neural network operates implicitly with data, while the Fuzzy model administers explicit information. Instead of boolean logic, a Fuzzy Expert System works with a variety of Fuzzy membership functions and rules for any kind of data [31]. The membership function scrutinizes the actual values generated during the experimental work. These values are worked on by the function in the Fuzzification subprocess. This process is essential to know the exactness of the truth for each premise of the rule. The inference process guides each rule in terms of the truth value, and then it is utilized to the conclusion part of each rule. Knowledgebase guides FES. The transformation to a crisp number of the fuzzy output is called defuzzification. Figure 11 depicts the structure of an FES.

The fuzzy-based approach has proven to be one of the useful tools with the ability to solve complex problems for spark ignition (SI) engines [13]. Saraswati et al. [32] developed and designed a system that worked together as a neural network. They used a fuzzy logic control system for an SI engine advance control, to maximize the brake torque timings of an SI engine. The fuzzy logic controller maintains a peak pressure of approximately 16^0 after the TDC. The fuzzy logic controller is the perfect match with the behavior of the SI engine in terms of the nonlinear nature.

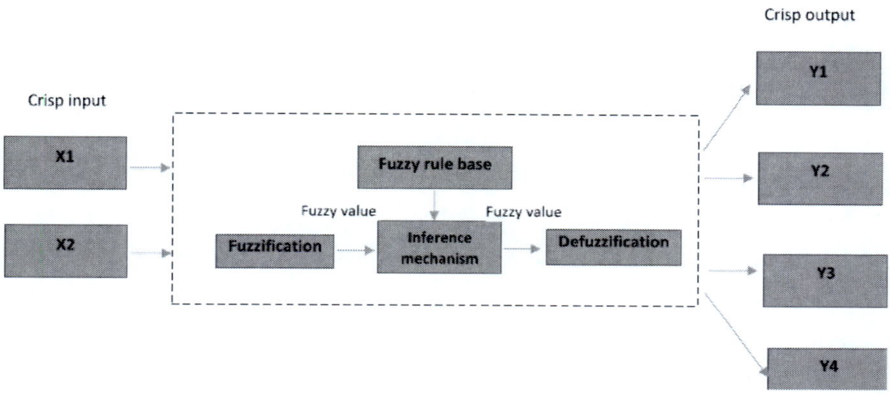

Figure 11. Structure of FES [14].

With all coordination and proper working, the recurrent network model labored with the controller to make out for the pressure existence with correct values in the cylinder. The neural network system employed by the researcher used spark advance, intake manifold pressure, crankshaft speed fluctuation, and equivalence ratio as the input to validate the experimental data. The fuzzy logic controller operates on the feedback signal received and provides the exact knowhow of the spot of peak pressure occurrence and its subsequent value. The existence of external turmoil was unable to create a dent in the working standards of the controller, which displayed fitting results as the researchers concluded it. Mozaffari et al. [33] anticipated the problems incurred during the cold starting of a spark ignition (SI) engine. So he devised an intelligent framework to make out for the temperature of the gases at the exhaust and HC emissions during the cold starting of the engine. Various software computing approaches provide the best model and frame network design to solve all complex and non-linear problems. Neuro-fuzzy computing, fuzzy controller, and swarm intelligent computing are the various software computing techniques. A hybrid learning scheme trained the anthology of the network under which each neuro-fuzzy sub-component resided. These methods are capable of providing accurate and reliable results for the problem, which was identified by the researchers.

Minimizing the pollutant emissions from a small spark ignition (SI) engine by proper utilization of the modeled fuzzy control system became the motive to research by Lee et al. [34]. The Fuzzy algorithm measures the necessary fuel required using the engine speed and the manifold air pressure. During this procedure, a fuzzy design environment was created, consisting of a fuzzy set editor and a fuzzy rule editor. Parameters transport themselves to the fuzzy inference kernel so extracted from the fuzzy development environment. Fuzzy inference kernel knowledge finds its place as a module in C++. In terms of fuel management, the control system showed reliability and gave precise solutions compared to the carburetor-based model. A dramatic decrease in carbon monoxide and hydrocarbon emissions were critical results of this study.

CONCLUSION

Scientists and researchers have used different simulation techniques to predict an SI engine's output and pollutant emanation using alcohol-gasoline blends. There is a place in this paper for various soft computing techniques that SI engines adopt. ANN is the most effective tool in simulating the exact results of the experimental value. A feedforward propagation algorithm is the most popular network system with LM and SCG as a training feature. The anticipated outcome for the correlation coefficient is very near to 0.9, and the RMS value is quite small. In terms of predictions, FES and ANFIS perform better than ANN. Also, RSM and SVM have been beneficial for the optimization process as they are very supportive in simulating different parameters. Modeling techniques saves much time, and the capital investment involved is comparatively meager.

Modeling techniques are quick and steadfast, so they are capable of replacing conventional experiments.

REFERENCES

[1] KianiD.M., Ghobadian B., Tavakoli T., Nikbakht M.A., NajafiG. Application of artificial neural networks for the prediction of performance and exhaust emissions in SI engine using ethanol-gasoline blends; *Energy* 35 (2010) 65–69.

[2] Najafi G., Ghobadian B., Moosavian A., Yusaf T., Mamat R, Kettner M., Azmi W.H. SVM and ANFIS for prediction of performance and exhaust emission of an SI engine with gasoline–ethanol-blended fuels; *Applied Thermal Engineering* 95 (2016) 186–203.

[3] Najafi G., Ghobadian B., YusafT., Mamat S. Optimization of performance and exhaust emission parameters of an SI(spark ignition) engine with gasoline-ethanol blended fuels using response surface methodology; *Energy* 90 (2015) 1815-1829.

[4] Mehrnoosh D., Asghar H., AsgharM. Thermodynamic model for prediction of performance and emission characteristics of SI engine fuelled by gasoline and natural gas with experimental verification; *Journal of Mechanical Science and Technology* 26 (7) (2012) 2213~2225.

[5] Tasdemir S., SaritaS.I., Ciniviz M., AllahverdiN. Artificial neural network and fuzzy expert system comparison for prediction of performance and emission parameters on a gasoline engine; *Expert Systems with Applications* 38 (2011) 13912–13923.

[6] Dagaut P., Togbe.C. Experimental and Modeling Study of the Kinetics of Oxidation of Ethanol-Gasoline Surrogate Mixtures (E85 Surrogate) in a Jet-Stirred Reactor; *Energy & Fuels* 2008, 22, 3499–3505.

[7] Yusri I.M., Mamat R., Azmi W.H., Omar A.I., Obed M.A.,Shaiful A. Application of response surface methodology in optimization of

performance and exhaust emissions of secondary butyl alcohol-gasoline blends in SI engine; *Energy Conversion and Management* 133 (2017) 178–195.

[8] Yusri I.M., Majeed B.A., Mamata R., Ghazalia M.F., Awada O., Azmi.W.H. A review on the application of response surface method and artificial neural network in engine performance and exhaust emissions characteristics in alternative fuel; *Renewable and Sustainable Energy Reviews* 90 (2018) 665–686.

[9] Awada O., Mamata R., Alic O., Azmia W.H., Kadirgamaa K., Yusria I.M., Lemand A.M., Yusaf.T. Response surface methodology (RSM) based multi-objective optimization of fusel oil -gasoline blends at different water content in SI engine; *Energy Conversion and Management* 150 (2017) 222–241.

[10] Liu Z., Zuo Q., Wu G., LiY. An artificial neural network developed for predicting performance and emissions of a spark-ignition engine fueled with butanol–gasoline blends; *Advances in Mechanical Engineering* 2018, Vol. 10(1) 1–11.

[11] Isin O., UzunsoyE. Predicting the Exhaust Emissions of a Spark Ignition Engine Using Adaptive Neuro-Fuzzy Inference System; *Arab J Sci Eng* (2013) 38:3485–3493.

[12] Zuo Q., Zhu X., Liu Z., Zhang J., Wu G., Li Y. Prediction of the Performance and Emissions of a Spark-Ignition Engine Fueled with Butanol-Gasoline Blends Based on Support Vector Regression; *Environmental Progress & Sustainable Energy* (Vol.38, No.3) DOI 10.1002/ep.

[13] Shrivastava N., Khan Z. Application of Soft Computing in the Field of Internal Combustion Engines: A Review; *Arch Computat Methods Eng* (2018) 25:707–726.

[14] Tekin M., Saridemir S. Prediction of engine performance and exhaust emissions with different proportions of ethanol-gasoline blends using artificial neural networks; *International journal of ambient energy,* 2017, 0143-0750.

[15] Abdallah N.A., Taob H., Bagaberc S., Alid O., Kamile M., Maf X., Awad O. Prediction of emissions and performance of a gasoline

engine running with fusel oil–gasoline blends using response surface methodology; *Fuel* 253 (2019) 1–14.

[16] Wonga P., Vongb C., Wonga H., Li K. Modelling and Prediction of Spark-ignition Engine Power Performance Using Incremental Least Squares Support Vector Machines; *AIP Conference Proceedings* 1233, 179 (2010).

[17] Thakur A.K., Mer K., Kaviti A. An artificial neural network approach to predict the performance and exhaust emissions of a gasoline engine using ethanol-gasoline blended fuels; *Biofuels,* 2017, 1759-7269.

[18] Bilgehan M. A comparative study for the concrete compressive strength estimation using neural network and neuro-fuzzy modeling approaches; *Nondestructive Testing and Evaluation,* Vol. 26, No. 1, March 2011, 35–55, Taylor & Francis, 1058-9759.

[19] Najafi G., Ghobadian B., Tavakoli T., Buttsworth D.R., Yusaf T.F., Faizollahnejad M. Performance and exhaust emissions of a gasoline engine with ethanol-blended gasoline fuels using artificial neural network. *Appl Energy* 2009; 86:630–9.

[20] Çay Y., Korkmaz I., Çiçek A., Kara F. Prediction of engine performance and exhaust emissions for gasoline and methanol using artificial neural network. *Energy* 2013;50:177–86.

[21] Kiani M.K.D., Ghobadian B., Ommi F., Najafi G., Yusaf T. Artificial neural networks approach for the prediction of the thermal balance of SI engine using ethanol-gasoline blends. In: *Proceedings of the international conference on availability, reliability, and security.* Springer, 2012. p. 31–43.

[22] Kapusuz M., Ozcan H., Yamin J.A. Research of performance on a spark ignition engine fueled by alcohol–gasoline blends using artificial neural networks. *Appl Therm Eng* 2015;91:525–34.

[23] Awad O.I., Mamat R., Ali O.M., Azmi W., Kadirgama K., Yusri I. et al. Response surface methodology (RSM) based multi-objective optimization of fusel oil-gasoline blends at different water content in SI engine. *Energy Convers Manag* 2017;150:222–41.

[24] Chen Y.-L., Chen S., Tsai J.-M., Tsai C.-Y., Fang H.-H., Yang I.-C., et al. Optimization of suitable ethanol blend ratio for motorcycle engine using response surface method. *J Environ Sci Health Part A* 2012;47:101–8.

[25] Yusri I.M., Mamat R., Azmi W.H., Najafi G., Sidik N.A.C., Awad O.I. Application of response surface methodology in optimization of performance and exhaust emissions of secondary butyl alcohol-gasoline blends in SI engine. *Energy Convers Manag* 2017; 133:178 95.

[26] Yusri IM, Mamat R, Azmi WH, Najafi G, Sidik NAC, Awad OI. Experimental investigation of combustion, emissions, and thermal balance of secondary butyl alcohol- gasoline blends in a spark-ignition engine. *Energy Convers Manag* 2016; 123:1–14.

[27] Najafi G., Ghobadian B., Yusaf T., Ardebili S.M.S., Mamat R. Optimization of performance and exhaust emission parameters of an SI (spark ignition) engine with gasoline– ethanol-blended fuels using response surface methodology. *Energy* 2015; 90:1815–29.

[28] Kalogirou S.A. (2003) Artificial intelligence for the modeling and control of combustion processes: a review. *Prog Energy Combust Sci* doi: 10.1016/S0360-1285(03)00058-3.

[29] Lee S.H., Howlett R.J., Walters S.D., et al. (2007) An modeling and control of internal combustion engines using intelligent. *Cybern System* 38:5–6:509–533.

[30] Kilagiz Y., Baran A., Yildiz Z., Murat C. (2005) A fuzzy diagnosis and advice system for optimization of emissions and fuel consumption. *Expert Syst Appl* 28:305–311.

[31] Tasdemir S., Saritas I., Ciniviz M., Allahverdi N. (2011) Artificial neural network and fuzzy expert system comparison for prediction of performance and emission parameters on a gasoline engine. *Expert Syst Appl* 38:13912–13923.

[32] Saraswati S., Agarwal P.K., Chand S. (2011) Neural networks and fuzzy logic-based spark advance control of SI engines. *Expert Syst Appl* 38:6916–6925.

[33] Mozaffari A., Azad N.L. (2015) An ensemble neuro-fuzzy radial basis network with self-adaptive swarm-based supervisor and negative correlation for modeling automotive engine cold-start hydrocarbon emissions: A soft solution to a crucial automotive problem. *Appl Soft Computation J* 32:449–467.

[34] Lee S.H., Howlett R.J., Walters S.D. (2004) Emission reduction for a small gasoline engine using fuzzy control. *IFAC Symp Adv Automot Control 200*–205.

In: Applied Soft Computing Techniques … ISBN: 978-1-53618-180-7
Editors: Amit Kumar Thakur et al.

Chapter 7

Energy Conservation Techniques for Food Processing and Manufacturing Industries

***V. Shinduja*[1] *and Gopinath Shanmugaraj*[2,*]**
[1]Department of Food Technology, Anna University, Chennai, Tamilnadu, India
[2]Aerospace Engineering, School of Mechanical Engineering, Lovely Professional University, Phagwara, Punjab, India

Abstract

This paper presents the overall review of the different studies that explore energy-saving opportunities in the food industry and also describes several case studies carried out in various food processing and manufacturing industries in different regions of the world. In all the literature first, the analysis of energy consumption in every processing of the food industry was determined. Based on the energy consumption, most efficient energy-saving practices were implemented, such as energy-efficient processing methods, improved equipment design, and proposed

* Corresponding Author's Email: shanmugagopi@yahoo.com.

energy-efficient numerical modeling to conserve energy in the food processing and manufacturing industries. From this study, it is known that the significant energy consumption in the food industry is due to electricity, heat, and cooling process. This excess energy intake can be reduced by implementing good energy-saving practices, which is described in this paper.

Keywords: energy conservation, food manufacturing practices, food processing, energy policy, energy consumption

INTRODUCTION

Most of the food that we are consuming nowadays is processed foods. The primary reason for processing the food is to increase its shelf life, convenient for easy preparation, reduction of food spoilage by the production of edible foods, new food product development for the newer generation, and the demand for fresh foods. To produce this kind of food, the food processors require energy sources. The essential forms of energy used in such types of industries are heat, mechanical machines, and fossil fuels.

The food industry is one of the higher energy-consuming sectors, but when comparing with the total production cost, it is considered to be not-energy-intensive. Only 3% of energy consumption is taking part in the whole production process [1, 2]. Due to the massive number of food industries running all over the world, it is considered to be a critical energy-intensive industry. In the US food industry, 4.4% of energy consumption was recorded in the year of 1998. From some survey reports, it came to know that the food industry is the fifth-highest energy-consuming industry out of 20 various biggest industries [3], and it is presented in Figure 1. So efficient energy management programs should be implemented to achieve the potential energy savings. It should be considered seriously, and severe actions must be taken to implement this from one production area to others [4].

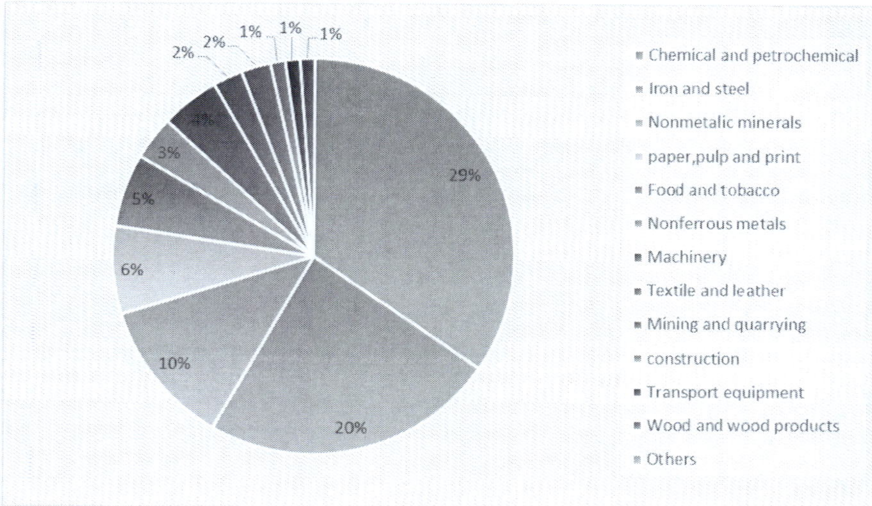

Figure 1. Distribution of various industrial energy use in 2005 [5].

In the food manufacturing and processing industries, most of the studies focus mainly on the determination of energy consumption. In 1976 the federal energy administration gave the full report of energy consumption list of US food industries. Out of total energy consumption in US 2.9% of energy was utilized by farm productions, 4.8% of energy was being used by the food processing and manufacturing industries, home food preparations used 4.3% of energy, 2.8% of energy was being used by out of home food preparations, wholesale food tradings used 0.5% of power, and retail food tradings used 0.8% of energy.

The report, as mentioned above of energy values, includes direct, indirect, capital, and transportation costs. In the food processing and manufacturing industry, only 18% of energy is used for the production of foods remaining 82% of energy is used for preparation for consumption, processing of foods, and marketing. The food processing segment uses the largest share of that 82 percent of the energy consumed. So energy conservation in food industries has to be intensely focused nowadays. In this review paper, the energy accounting methods, which are an essential component of energy conservation technology, is discussed.

ENERGY CONSUMPTION

Food processing involves transforming harvested crops, slaughtered animals or meat, milk or milk products, fish or seafood, and other raw food supplies into edible food products that are attractive and often having long shelf-life. Such industries include processing methods such as different types of cooking, including baking, smoking, precooking, steaming, and roasting. It also includes non-cooking techniques such as drying, fermentation, and canning. Apart from these processing techniques processing industry includes packaging, storage, distribution, and transportation of final products to the customers and transportation of raw materials to the factory [6].

The food industry can be classified into specific industry sectors depending on different food processing methods and manufacturing techniques they use: fermented products manufacturing, instant or ready-to-eat food manufacturing, candied fruit manufacturing, processed or standardized or enriched milk and dairy-based food manufacturing, confectionery, bakery, chocolate, condiment and canning industry [7]. Various bio-transformation, separation processes, and chemical reactions are involved in this processing and manufacturing methods [8]. Food processing practices, such as cooling, drying, and sterilizing, help reduce damages during food delivery, packaging, and handling. The introduction of such a mechanism to achieve more efficient processing capacities has, in the last several decades, forced the food manufacturing industry to be contingent heavier on energy. The majority of those energies come from petroleum products [9].

In recent days, other manufacturing technology developments that are not related to the food industry are influencing the food manufacturing and processing methods significantly. One such system is automation, and it has generated in the industry. Various processing methods have become automated, such as washing raw fruits & vegetables, canning, filling, sealing, and labeling. Energy-consuming machines, therefore, replace manual labor skills. This will ensure that the food manufacturing and processing industries will see a substantial increase in power consumption in the future [7].

MAJOR ENERGY SOURCES

In various processing segments, industries generally use different energy sources such as charcoal or wood, natural coal, petroleum or oil-based, natural gas, biofuel, solar, electric, wind, and nuclear. Significant energy is required to continuously operate various appliances in the food manufacturing and processing industries such as air conditioning systems, dryers, refrigerators, electric motors, boilers, gas-fired ovens, heating, and ventilation. Out of all energy sources, electricity is the primary type of energy use in the food industry. Two-thirds of its consumption is used for the operation of industrial machinery like factory lighting, compressors, boilers, pumps, blowers, sorting, and packaging conveyors. The critical area of ensuring efficient energy use in the food industry is the accurate and adequate operation of several processes. The most potent key to energy savings in plant operations is the optimization of food processing techniques.

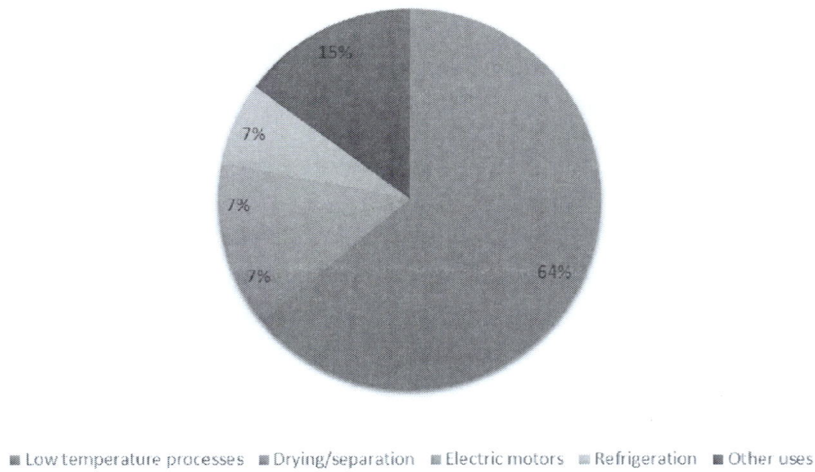

Figure 2. Breakdown of the food sector energy use [11].

The use of energy consumption varies depending on the different processing technology used in food processing. Low-temperature processes, for example, use 64 percent of the total energy, while motors, drying or separation, and refrigeration use only 7 percent of the total energy consumed [11]; Figure 2 illustrates the split-up of energy use in the food industry.

METHODS TO REDUCE ENERGY CONSUMPTION

Various applications of appliances such as room ventilation systems, heating, engines, compressors, furnaces, air conditioning, and boilers in the food processing industries require energy in various forms. The appliances mentioned above should be used effectively to reduce energy use and CO_2 emissions [12]. Each industrial process is unique, and the energy efficiency problem of that process has a specific area to improve, which involves a thorough understanding of the entire process. Several growing sectors have the potential for energy efficiency upgrades. Industrial plants must be built with an energy conversion device of at least one type. Also, it is essential to remember that the equipment is deemed to be practical, durable, and to have environmental impacts [13].

Research by Utlu and hepbasli[14] identifies the significant areas that waste energy in the food production and processing industries. The essential sources are water treatment plant, room environment controlling & conditioning unit, air conditioning and filter equipment, Internal heating ventilation systems and boiler plants, air distribution system, lack of sun-roofing installation or maintenance, and extensive use of artificial lighting, leaks through doors or curtains. Energy saving is essential in the food manufacturing and processing industry as a result of the below-listed reasons.

- Primary sources of energy that are used in the food industry are nonrenewable; this will affect the availability of energy in future needs.

- Cost-efficient energy usage methodology in the food industry can save up to 30 percent of total energy consumption.
- Environmental pollution can be reduced by following energy-saving practices in the food manufacturing and processing industry.
- Investing in energy-saving practices in the food processing industry can provide a better return on investment than newly investing in energy supply.

Ogunjuyigbe et al. [15] carried out an energy audit survey for Nigeria's particular beverage and food industries. The author found from this survey report that the total energy consumption of approximately 1MWh can be saved by replacing the conventional motors with lates less energy-consuming engines and replacing the lights used in the industry with LED lights and fluorescent light lamps.

The author focuses on implementing the energy-saving practices in the following systems, which are mainly used in Nigerian food processing industries, from the energy audit report.

Air Conditioning, Heating and Ventilation System

Air conditioning, heating, and ventilation system are some of the most critical energies utilizing equipment in Nigeria food processing industries. In those industries, it was found that it dissipates heat, and also it was much dusty. To reduce the energy usage in such equipment's proper and regular cleaning must be done. The temperature of the air conditioning system should be appropriately set to avoid wastage. The doors present in the air-conditioned room must be adequately fitted with springs to allow for self-control to trap cold air. The ventilation system should be installed to provide a constant temperature of fresh air.

Electric Drives

Electric drives like electric motors, compressors, refrigerators, fans, pumps consume electrical energy, about 65% in food processing industries. Improving the energy-saving practices in this equipment will significantly reduce the electricity consumption in these industries. Electric motors used in most of the Nigerian food processing industries are rewound more than twice because of that they produce more heat, and also they are weak. To prevent such kind of energy wastage, old motors must be replaced with new engines. Proper maintenance of these electric devices helps to reduce energy waste. Electric motors can be reconfigured to obtain at least 33 percent of the rated output from delta to star connection. Installing the electrical drives on electrical motors at variable speed and with soft start methods can also aid in energy-saving practices.

Electric Heaters

Typical electrical heaters used in food manufacturing and processing industries are electric salt bath furnaces, induction, direct arc electrical furnaces, infrared heaters, high-frequency resistance heaters, dielectric heaters, and resistance heaters (ovens, boilers, metal sheaths, furnaces, etc.). In Nigeria, more heat loss in the steam-generators occurs in the food industries because of improper design layout of pipes carrying the heated steam from boilers to the processing unit. By replacing the existing boilers with a new one, and also using small size boilers, this type of heat loss can be reduced. High-precision control burners have to be installed to manage the continuous and accurate air-fuel ratio.

Lights

According to the survey taken in Nigeria, lighting systems of food processing industries are not adequately controlled, and all the lighting points are always on even though it is day time. Energy-saving practices should be implemented in this lighting system. Through these practices, we can save plenty of electric energy in food processing industries. A higher efficiency lighting device replaces lower efficiency lighting systems. Walls and ceilings of the food processing industries can be painted with light colors or pure white. To increase the brightness of the workplace, light reflectance can be used. More amount of energy can be saved through these processes.

Compressors

In food processing industries, compressors play a critical role in air compression, which results in 17% of global industrial electricity usage. This can be reduced by using VSD compressors, reducing the air leakages, reducing the air pressure from 7 to 6 bars.

CASE STUDIES

Wang Hee Lee and Martin R. Okos [16] have done a study to evaluate energy and water consumption in the three primary food processing industries. They are the edible bean processing industry, dairy industry, and corn masa processing industry. In this study, the author proposes alternatives to reduce energy and water consumption in the food mentioned above sectors. The main focus of their research is to determine energy consumption by doing the audit data collection on a plant scale. To do laboratory-scale experiments to check the quality of the product by reducing the supply of water and energy to propose a computer-based simulation for the development of the new system, which reduces the usage of water and energy in the food industry. From this study, they found that the water recycling method will decrease water usage by 90%, and also it will reduce water usage by up to 55% when compared to the traditional way. In the water recycling method, 70% of energy consumption was reduced, which is required for heating. When this method is scaled up for the plant scale process, water usage will be reduced to 95%, and the energy consumption is reduced by about 80%.

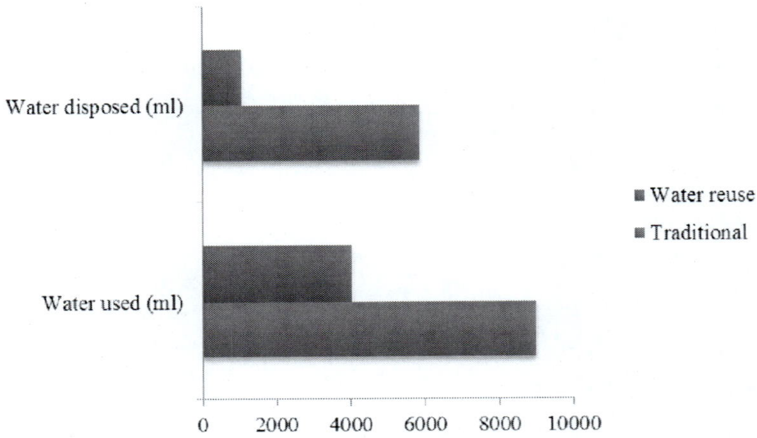

Figure 3. Water utilization and discharge range comparison between traditional and water recycling corn masa processes [16].

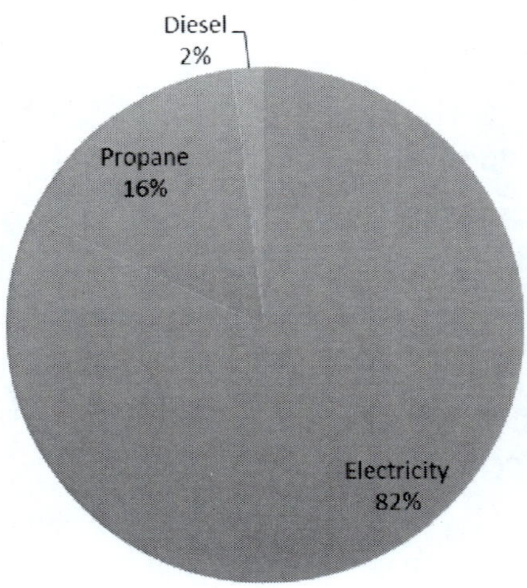

Figure 4. Distribution of various types of energy consumption in sausage factories [17].

In the processing of 3kg dry corn, they implemented the water reuse process; and the result shows that wastewater is reduced by 88%, and the usage of water during the corn processing was reduced by 55%. The total amount of energy used for the water recycling method was calculated based on the temperature utilized, and the result shows that energy input for heating was reduced by 70%.

The meat processing industry is one of the largest energy-consuming sectors in the world. The process of drying and aging of cheese, ham, and sausages consumes a lot of energy in the meat industry. Junes et al. [17] suggest to use the extreme atmospheric conditions for drying can help to reduce energy consumption by 25%. Apart from this, 50% of energy can be saved by using absorbent materials inside the processing room. They also have done a case study regarding the energy consumption of 20 sausage industries in Portugal. They reviewed energy consumption, energy-efficiency measures, and energy indicators in all 20 sausage industries in this research. The result of this case study shows that energy consumption is mainly due to electricity (82 percent) and fuel on a lesser scale (18 percent) in this type of food processing industry. Sausage processing involves mostly cold storage and meat drying. So the primary consumption of electricity is due to the refrigeration system and dryers. By implementing simple practice measures in the refrigeration system (control settings, defrost optimization, improved door protection, and repairs) used in this sausage industry, it can be reduced to up to 24 percent. Ortega et al. [21] proposes a semi-automated method to monitor the humidity and air conditioning inside the rooms and automatically control the conditioning unit will save 50% of the electrical power consumption.

Nathan Pelletier et al. [18] conducted a review of energy consumption in 20 various food processing and manufacturing industries in the United States. While comparing energy consumption among the food industries surveyed, the wet corn milling industry consumes more than 15% of the total energy consumption of the food industry terms. It is the most energy-intensive industry among the food processing industries. In energy consumption, the sugar beet industry takes second place (7 percent). At the same time, malt beverage, soybean oil mills, and meatpacking plants use about 5 percent of total energy in this sector. They also determine which single process is energy-intensive in this study. The report says the process of heating and cooling in the food industry is the most energy-intensive and accounts for 44.6 percent of total energy consumption. For example, the most energy-consuming cycle of dried milk powder processing in the dairy industry; this can be seen in Figure 5.

From this review, it is found that the dairy industry has a higher chance of energy conservation by implementing the following techniques. By increasing the use of HTST regeneration (high-temperature short time) process, building ad steam line insulation, and conservation practices during sanitation, the steam requirement can be reduced to about 32 percent. By implementing efficient motors, lighting, and insulation, this food industry can save 23 percent of electricity.

Through this survey, they concluded that cooling/refrigeration, heating, and boiler losses were the most energy-consuming processes in the food processing and manufacturing industries, representing 29.1 percent, 15.5 percent, and 22 percent of total energy intake, respectively. Energy-saving techniques such as recuperating heat from compressors, recovering and reusing hot and cold streams, and using more efficient boilers can lead to increased energy efficiency. For example, their survey result shows that in the soybean oil mill industry, recycling of the dryer air can lead to 25% energy saving in the dryers, and also using the stack gasses for the heating of boiler feed water can reduce the boiler losses by 37%. Implementation of efficient lighting systems and motors can also lead to conserving energy use, about 23% in the food processing and manufacturing industries.

A. Almena et al. [19] propose a model-based optimization condition for double drum – dryer (under atmospheric conditions), which is used in the manufacturing of breakfast cereal porridge. This research aims to reduce energy consumption, particularly in drying equipment, which is used in most of the food manufacturing and processing industries. In this study, the model minimizes the energy demand of the dryer based on the optimal steam temperature and optimal rotation speed. This work demonstrates the potential of model-based approaches to design energy-efficient industrial dryers for the food sector. Thus it will achieve the goal of energy saving in food manufacturing and processing industries. To develop an energy-efficient drum dryer, two main variables were manipulated in this study, and they are input steam temperature T_{steam} (°C) and the rotation speed of the drums ω (rpm). By optimizing the optimal steam temperature and drum rotating velocity, the overall energy required to dry the slurry can be minimized. Apart from these different process variables ate assessed, they are initial and final water content, sheet thickness, and flour composition. From this result, there is a potential for further reduction of energy by adjusting the initial and final water content in the product formulation. The result reveals that thicker slurries require higher steam temperature and longer residence time, whereas thinner slurries require lower or medium temperature and shorter residence time.

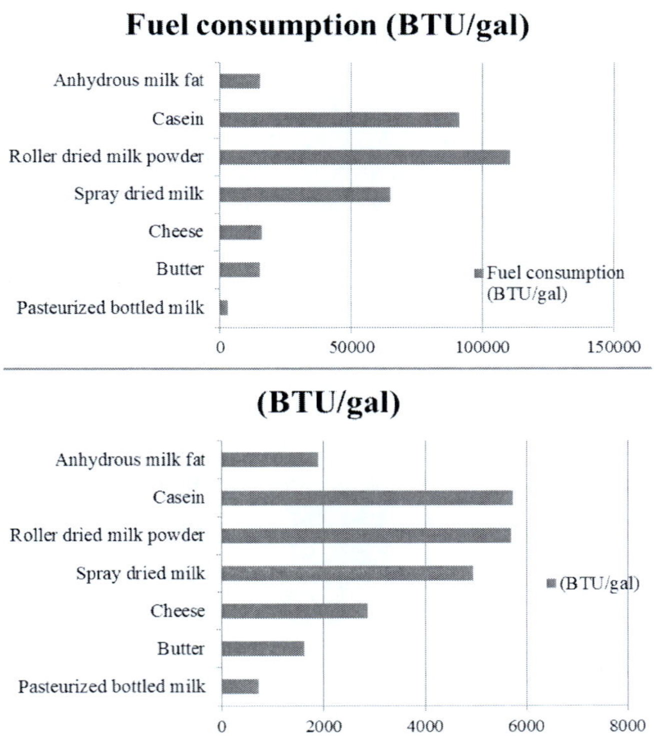

Figure 5. Fuel and electricity consumption range in the production of various dairy products [18].

James C Atuonwu et al. [20] conducted a comparison study on the dehumidification drying energy efficiency versus conventional convective drying techniques. The effect of dehumidification dryers such as the method of adsorption, the condensation form and heat pump dryers, and the impact of traditional convective dryers were analyzed in this research, and their energy efficiency was compared. Based on the mathematical models, the energy efficiency of every dryer is determined. The results of this study concluded that dehumidification of the drying air has a higher energy consumption advantage compared to conventional convective drying techniques.

Fermentation is an essential part of the food manufacturing process. To support the fermentation process, a suitable environment should be provided. The significant energy consumption for the fermentation process becomes of the air conditioning system. Jin B et al. [22] discussed an energy-saving strategy of an air-conditioning system of a tea brick fermentation room under different environmental conditions. They quantitatively calculated the indoor moisture and heat load of the tea brick during the fermentation process. They have modeled a porous media using a moisture and heat coupling transfer theory. COMSOL multiphysics was used to solve the problem numerically. Analysis has been done for the fermentation cycle to find out the excess loss of energy in the air-conditioning system. They have compared the results of the circulating air system (AS) and one return air conditioning system (ORCS). It is concluded that the ORCS performs 30% to 55% more efficient if the residual heat is inside the chamber is zero. But, CAS shows 16 to 22% more efficient than the ORCS when the residual heat is more than the zero.

Magdalena et al. [23] studied the effects of ultrasound drying kinetics, blanching, and other pretreatments on the energy consumption of drying parsley leaves. Interestingly, they have found that energy consumption can be reduced up to 33% if ultrasound is used in place of commercial steam blanching. The color retention is also found to be better for the ultrasound case. The process also proved to be faster when compared with conventional methods.

CONCLUSION

Since the food manufacturing and processing industry is the world's fifth-largest energy-consuming industry in the overall manufacturing sector, it is essential to reduce energy utilization in these industries. In this review paper, most of the significant energy-saving practices which can be followed in the food sector are discussed. The above mentioned energy-saving practices must be followed in the food industry to reduce energy consumption rapidly. There also many probabilities for further refining the energy efficiency in the food sector through process analysis and improvement and implementation of effective governance strategies and new energy policies. Future studies to advance the energy efficiency in the food industry must focus on the growth of existing plants, development of energy-efficient process technology, design and optimization of energy-efficient food processing equipment and additional research about zero discharge plant.

REFERENCES

[1] De Groo, Henri, Erik verhoef and peter Nijkamp. 2001. "Energy savings by firms: decision-making, barriers, and policies, Energy Economics." *Tinbergen Institute Discussion papers* 23:717–40.

[2] Sandberg, P., and M. Soderstrom. 2003. "Industrial energy efficiency: the need for investment decision support from a manager perspective" *Energy Policy*. 31:1623–34.

[3] Energy Information Administration, *Annual Energy Review 2003*, Report no. DOE/EIA-0384, http://tonto.eia.doe.gov/bookshelf/.

[4] Damien C. A. Muller, François M. A. Marechal, Thomas Wolewinski, and Pieter J. Roux. 2007. "An energy management method for the food industry." *Applied thermal engineering* 27: 2677-86.

[5] Data based on International Energy Agency, *World Energy Outlook 2008*, https://www.iea.org/media/weowebsite/2008-1994/WEO2008.pdf.

[6] National Training Agent (NTA). 2015. *Quantitative research on labour market demands in the food and beverage manufacturing industry*, Trinidad, pp. 1–44.

[7] Qi J., Zhao J., Li W., Peng X., Wu B., and Wang H. 2016. "The circular economy-oriented practice in the food manufacturing industry." *Development of Circular Economy in China* 201–222. Accessed October 13, 2016. doi:10.1007/978-981-10-2466-5_10.

[8] Arora, A. et al., 2014. "Indian Food & Beverage Sector: the new wave. Food & Beverage Technology." *5th International Exhibition & Conference*, Grant Thornton, India.

[9] Waide, P., and C. Brunner. 2011. "Energy-Efficiency Policy Opportunities for Electric Motor-Driven Systems," *IEA Energy Papers*, doi: 10.1787/5kgg52gb9gjd-en.

[10] Aderemi, A. Ilori, M. Aderemi, Helen and Akinbami, John-Felix. 2009. "Assessment of electrical energy use efficiency in Nigeria food industry" *African Journal of Food Science*. 3: 206-16.

[11] Law, Richard, Adam Harvey and David Reay. 2013. "Opportunities for low-grade heat recovery in the UK food processing industry." *Applied Thermal Engineering*. 53: 188–96.

[12] Ali Boharb, Amine Allouhi, Rahman Saidur, Tarik Kousksou, and Abdelmajid Jamil. 2017. "Energy conservation potential of an energy audit within the pulp and paper industry in Morocco." *Journal of Cleaner Production*. 149: 569–81.

[13] Mossie, A. T. 2016. "Energy utilization assessment in Ethiopian industries (case study at Mughar Cement Factory (MCF))." *International Journal of Engineering Research & Technology*. 5: 189–93.

[14] Utlu, Z. and Hepbasli, A. 2009. "Exergo economic aspects of sectoral energy utilization for Turkish industrial sector and their impact on energy policies." *Energy Policy*. 37: 577–87.

[15] Ogunjuyigbe, A. S. O., T. R. Ayodele and S. M. Ogunmuyiwa. 2015. "Improving electrical energy utilization in some selected Nigerian food and beverage industries." *Sustainable Energy Technologies and Assessments.* 12: 38–45.

[16] Lee, Wang Hee and Martin R. Okos. 2011. Sustainable food processing systems – path to a zerod is charge: reduction of water and energy. *Procedia food science* 1: 1768 - 77.

[17] Nunes, J., Pedro D. Silva, L. P. Andrade and Pedro D. Gaspar. 2016. "Key points on the energy sustainable development of the food industry – A case study of the Portuguese sausages industry." *Renewable and sustainable energy reviews.* 57: 393 – 411.

[18] Pelletier, Nathan, Eric Audsley et al. 2011. Energy Intensity of Agriculture and Food Systems. *Annual Review of Environment and Resources.* 36: 223-46.

[19] Almena, K. R. Goode, S. Bakalis, P. J. Fryer and E. Lopez-Quiroga. 2019. "Optimizing food dehydration processes: energy-efficient drum-dryer operation. *Energy procedia.* 161: 174-81.

[20] Atuonwu, James C., Xin Jin, Gerrit van Straten, Henk C van Deventer Antonius and J. B. van Boxtel. 2011. "Reducing energy consumption in food drying: opportunities in desiccant adsorption and other dehumidification strategies." *Procedia food science.* 1: 1799- 805.

[21] Alcazar-Ortega, Manuel, Escrivá Guillermo, Alvarez C. and Domijan, Alex. 2011. "Active Demand Response Strategies to Improve Energy Efficiency in the Meat Industry." *Proceedings of the world renewable energy congress. Energy End-Use Efficiency Issues (EEE),* Linkoping, Sweden. doi: 10.3384/ecp11057976.

[22] Jin, Baohong, Xiaohong Nan, Xuhao Ning and Zanshe Wang. 2020. "Energy-saving operation strategy of the air conditioning system in tea brick fermentation room under different outdoor meteorological parameters." *Sustainable Cities and Society.* 53. Accessed October 14, 2019. doi: 10.1016/j.scs.2019.101883.

[23] Magdalena Sledz, Artur Wiktor, Katarzyna Rybak, Malgorzata Nowacka and Dorota Witrowa-Rajchert. 2016. "The impact of ultrasound and steam blanching pre-treatments on the drying kinetics, energy consumption and selected properties of parsley leaves." *Applied Acoustics*. 103: 148-56.

In: Applied Soft Computing Techniques … ISBN: 978-1-53618-180-7
Editors: Amit Kumar Thakur et al.

Chapter 8

EFFECT OF BIODIESEL ON PERFORMANCE, EMISSIONS AND COMBUSTION

Jayashri N. Nair[*]
Mechanical Engineering Department, VNR Vignana Jyothi Institute of Engineering and Technology, Hyderabad, Telangana, India

ABSTRACT

Globalization and industrial growth have contributed to the decline of conventional fuels since these fuels are not usually renewable. Soaring fuel demand can lead to the global fuel crisis, and this has urged researchers to perceive alternative sources that are sustainable, readily available, and, most importantly, environment friendly. Biodiesel has proved to be a technologically and commercially viable alternative to fossil fuel by a conglomeration of investigations. Biodiesel assists in curtailing waste and emissions with the application of biodegradable products, and thereby protecting greenhouses. Biodiesel is mainly made up of vegetable oils and can be used without alteration in any diesel vehicle. A mixture of biodiesel is used to attune the environment that could still have a significant greenhouse effect. This chapter analyzes and

[*] Corresponding Author's Emails: jayashri@vnrvjiet.in; jayashri.mtech@gmail.com.

presents a comprehensive report on biodiesel output and pollution from various feedstocks testing on multiple engines. Results show that, based on their sources, different chemical compounds of biodiesel have influenced their properties, performances, emissions, and combustion. Investigations results discern that biodiesel–diesel is characterized by shorter ignition delay a reduced heat release rate as well as a slightly higher efficiency at the cost of fuel consumption. Analyze show that biodiesel mixtures with small volume content can be used in current CI engines without significant modifications.

Keywords: methyl ester, performance, engine, emissions, combustion

INTRODUCTION

The demand for energy has increased exponentially with the society's socio-economic growth. Different industries, such as manufacturing, shipping, livestock, domestic, etc. require electricity (Mirza UK et al., 2008). There are various energy sources such as coal, gas, solar, nuclear, hydro, water, etc., but still, transport relies heavily on fossil fuels. The imperious dependency on oil products, as well as associated economic and environmental issues, has created a disquieting stage (Chang et al. 2003; Correa et al. 2008). With finite and concentrated reserves, petroleum declines at a breakneck pace. This is the driving force for researchers, scientists, and industries to look out and explore for substitute fuels for diesel engines.

Rapid depletion of fossil fuel is a matter of concern for the future generation. Natural fuels are scarce and also lead to greenhouse gas emissions. This prompted us to explore alternative fuels with similar properties to natural fuels. A great deal of work is currently underway around the world to find alternative fuels that can satisfy current and future energy demands without enhancing global warming impacts (Moka et al., 2014). Biodiesel has proven its potential over the years. It assists in providing fuel economy and energy security. One of the main reasons is that biodiesel can be used in internal combustion engines without significant modifications. It is also environmentally friendly, non-toxic, and biodegradable (Arbab, MI et al., 2013). Biodiesel can be injected, stored, and burnt in the same way as diesel. It can be defined as a mixture of sustainable lipid feedstock of long-chain fatty acid monoalkyl esters (FAME). Vegetable oil or animal fat can contain lipid stock. Biodiesel is typically made up of small chains esters, generally methanol or alkyl fatty acid ethanol. It is currently produced of non-edible oils. Biodiesel blends have claimed to diminish unburned hydrocarbons, carbon monoxide, and particulate matter. Nonetheless, NO_x emissions continue to be a concern.

SCENARIO IN INDIA

India's economy will grow at the fastest pace by 2040. According to the World Oil Outlook 2040 study by the Organization of Petroleum Exporting Countries (OPEC), the country's demand for oil would eventually rise. Developing countries led by India and China are projected to account for 95 percent of global primary energy production. In India and China, energy needs are expected to increase by about 22 million barrels of oil equivalent daily in India and 21 million barrels of oil equivalent from 2015 to 2040, according to the (OPEC) report. With an average annual growth rate of 6.5 percent during the 2015-2040 timeframe, India is believed to be the fastest-growing nation. According to a report published in New Delhi on 24 March 2009 by Greenpeace, renewable energy will meet more than 35% of India's demand for power by 2030. The Ministry of New and Renewable Energy (MNRE) has prepared the National Biofuel Policy. This body governs the use of renewable energy resources in the Indian context (Kumar et al., 2011). Biofuels, including bioethanol and biodiesel, are expected by MNRE to be combined with fossil mineral fuel by 20% by 2017. Oils are classified as edible and non-edible and edible oils such as groundnut, peanut, etc. are used primarily to satisfy the dietary needs. In India, no vegetable oils, including rapeseed and mustard, soya or palm oil, are used in the biodiesel industry because India is not the independent producer of edible oils. To respond to domestic demand, it relies heavily on imports of palm oil and vegetable oil. However, if trees are grown on land not threatening to crops on marginal or distributed land, there is no direct conflict with food security between the use of non-edible seed oils from forest trees and reserve (Dwivedi et al., 2014). It is essential to find feeds that are not edible and are suitable for biodiesel synthesis. Around two million tons per year of potential output of non-edible oils are expected in India. Sal, mahua, neem, Pongamia, and jatropha seeds are the most common sources of oil. Jatropha and Pongamia were identified as probable biodiesel feed inventories in India following extensive research.

PERFORMANCE TRENDS OF DIFFERENT BIODIESEL

Effect of Biodiesel on Brake Thermal Efficiency

Brake thermal efficiency is the output power determined in comparison to the calorific value of the fuel with the aid of the dynamometer. (Pandhare et al., 2013) dealt with methyl esters of Jatropha and its blends. Biodiesel were added to diesel from 5%, 10%, 20%, 40%, 60%, 80% and 100% by volume. Brake thermal efficiency of diesel was highest at all loads compared to the blends. The author also tested blends for different compression ratios. The efficiency of the Jatropha oil methyl esters blends increased with a rise in the compression ratio. Clearance volume inside the cylinder minimizes as the compression ratio rises. The temperature of the air-fuel mixture compressed in this small volume increases, which promotes better mixing and evaporation of fuel-air mixture. Increment in brake thermal efficiency was discerned as a result of more energy released due to the combustion of fuel. Instead of diesel (Nayak et al., 2014) used biodiesel additive (Dimethyl carbonate) for preparing different blends of Mahua biodiesel and tested for different compression ratios. With an increment of additives in biodiesel increase in brake thermal efficiency was noted. B85 blend curve approached diesel curve; however, diesel fuel depicted maximum efficiency at all loads.

On the contrary, in another study, the addition of additives increased the efficiencies of biodiesel than diesel. Multi-DM-32 diesel additive was added to jatropha biodiesel by (Rao. Y. H. et al., 2009). Three jatropha Methyl ester blends were made by adding methyl esters in different proportions to diesel. For comparison, 100% Diesel and 100% Jatropha Methyl ester were also tested. Results reported that there was no much difference between 100% Jatropha Methyl ester and its blends. In another study (Nair J.N. et al., 2017) reported higher brake thermal efficiencies of Neem biodiesel blends than diesel. As the amount of biodiesel decreased, the efficiency increased, but diesel showed the lowest brake thermal efficiency. An increase in brake thermal efficiencies of all neem biodiesel blends was attributed to the high oxygen content of Neem biodiesel. But pure Neem biodiesel has shown a reduction in brake thermal efficiency than diesel at all loads except at higher loads (shivalakshmi et al., 2011). (Bueno et al., 2017; Panwar et al., 2010) experimented with Castor oil methyl ester on a four-stroke cylinder direct injection engine, respectively. Lower blends of Castor methyl ester depicted higher brake thermal efficiency. Despite their weak spray related properties, it did not affect the thermal efficiency.

Effect of Biodiesel on Brake Specific Fuel Consumption

The efficiency of fuel consumption is reported by brake specific fuel consumption (BSFC), by burning of fuel and the torque produced. The majority of kinds of literature reported higher BSFC for biodiesels. Methyl Esters were prepared with palm oil, waste cooking oil, a mixture of palm oil, and waste cooking oil (Ndayishimiye et al., 2011). BSFC decreased with rising load for all samples. Tests showed higher BSFC for all blends in contrast to diesel. Almost a 17-25% increase in BSFC was noted with palm oil biodiesel and waste cooking oil. Similar readings were achieved by (Celiketan et al., 2012) with biodiesels of rapeseed and hazelnut. Diesel depicted the lowest BSFC. The author attributed an increase in BSFC of methyl esters to their low calorific value and high densities than neat diesel as a large mass of fuel is needed to produce similar torque as diesel. Contrast to these readings (Santos et al., 2013) experimented with biodiesel produced from peanut and soybean oil, blends of peanut biodiesel, clean diesel on small and large capacity engines for various speed. With a more modest engine decrease in BSFC was observed with all combinations at a lower speed, but BSFC slightly increased at higher loads. With the larger engine, BSFC increased at higher speed as at the higher speed of engine friction losses are high, which consumes more amount of fuel. However, for smaller engine B5 blend showed the least BSFC. When pure palm biodiesel and soybean biodiesel were tested separately against neat diesel, pure soybean methyl ester showed maximum BSFC, and pure diesel showed the least BSFC. B20 of rice bran biodiesel showed the least BSFC among all the blends, including diesel, as reported by (Umesh et al., 2014).

Combustion Trends of Different Biodiesel

Effect of Biodiesel on In-Cylinder Pressure

The magnitude of pressure developed inside an engine cylinder decides the torque and power output of an engine. So the timing, as well as the magnitude of In-line cylinder pressure during compression and expansion stroke, needs to be measured. In-line cylinder pressure rises steeply when a mixture of fuel and air in isolated areas auto-ignite even before it has not mixed expertly. This phenomenon is called premixing, which is followed by controlled combustion. Continuous injection of fuel into the cylinder assists inadequate mixing with air under controlled combustion. Used cooking oil biodiesel and its blends showed similar nature of curves as diesel at all load conditions (Rao et al., 2008). The higher percentage of biodiesel with higher viscosity values discerned elevated values than diesel. Retardation of injection timing can reduce the peak pressures of biodiesel. In such a study, biodiesels from rapeseed oil, soybean, and sunflower oils were tested along with standard diesel fuel to study their combustion phenomena (Wood et al., 2015). Pure Soybean biodiesel, Beef tallow biodiesel B50, and diesel were further tested to evaluate the difference in performances with the initial tests with previously mentioned samples. An increase in peak torque was noted for B50 biodiesel when the start of the injection timing was retarded in comparison to diesel. This also resulted in a decrease in the inline cylinder pressure. With the same injection timings, B50 Tallow and Soybean biodiesel showed less torque.

To compensate for lower calorific value with the same amount of torque as diesel, the fuel injection supply was varied. With this arrangement, B50 reported the most moderate inline cylinder pressure. By allowing more fuel injection in the cylinder, maximum torque can be induced before maximum cylinder pressure attainment. (Sahoo et al., 2009) conducted test with neat biodiesel from jatropha, Karanja, and polanga oils and their blends (20.0 and 50 .0 by v %) at varying loads (0, 50, and 100%) against reference fuel as diesel. For the biodiesels and their blends, the peak appeared slightly after TDC, which the author has justified as safe and efficient. It was postulated that had the peak appeared before TDC would have resulted in knocking, affecting its durability.

Effect of Biodiesel on Ignition Delay

The span between the time of ignition and burning of the combustible mixture is known as ignition delay. Compression temperature, air velocity, air temperature, injections, pressure, fuel temperature, and engine size are the factors that influence ignition delay. Most of the researchers have characterized biodiesel with the early start of combustion (SOC) and shorter ignition delay as compared to diesel. (Bittle et al.,2010) studied ignition delay with biodiesel(Name not specified) was at mid and high loads the combustion durations were less with biodiesel, which can be credited to the faster diffusion rate and faster burning rate of the fuel. Substantial combustion duration was noted at lower loads of engines due to shorter ignition delay. At smaller loads, the rate of diffusion is not high enough to burn the biodiesel at a higher rate. Similar reports were claimed by (Sinha et al., 2007) with rice bran methyl ester, where early combustion was accounted to advance injection timing.

Though biodiesels have higher viscosity and density than diesel, they have shorter ignition delay due to its complex and rapid preflame reaction at higher temperatures. Inline cylinder temperatures are very high during combustion. This leads to the thermal cracking of biodiesel, which eventually produces lighter compounds. These lighter compounds ignite eventuating in shorter ignition delay. Early start of combustion and shorter IDs account for higher cetane number (CN) of the biodiesel (Sánchez-Borroto et al., 2014). As the chain length of fatty acid composition increases, the cetane number of biodiesel also rises (Knothe et al., 2003). This also accelerates a reduction in NO_x formations where the nitrogen in the fuel is prevented from reacting before the combustion ends. Whereas (Shahabuddin et al., 2013) credited it to the lower compressibility of the biodiesel.

Effect of Biodiesel on Heat Release Rate

The heat release rate influences the efficiency of the cycle and maximum cylinder pressure. (Gummus et al., 2010) tested hazelnut biodiesel and its blend where modifications such as increasing of injection timing, compression ratio, and injection pressure, provided a substantial improvement in combustion and heat release characteristics. The combustion of biodiesel started early, and it became prominent with an increase in biodiesel content. Heat release during premixed combustion is higher for diesel because of its high volatility and longer ignition delay. (Gautam et al., 2013) stored cotton seed biodiesel and blends for about six months, and then its properties were tested. It was reported that the heat release rate for biodiesel blends was higher than the baseline diesel. In a similar study (B. Tesfa et al., 2013) discerned that ignition of the biodiesel started before the ignition of diesel owing to high bulk modulus, higher viscosity and higher cetane number of biodiesel.

EMISSION TRENDS OF DIFFERENT BIODIESEL

Engine exhaust is one of the primary reasons for increasing pollution in the environment. These exhaust gases are harmful and linked to the health problems of human beings. Exhaust gases contain carbon monoxide, carbon dioxide, nitrous oxides, particulates, water vapor, and numerous other contaminants.

Effect of Biodiesel on CO Emissions

(Abed et al., 2018) studied blends of transesterified waste cooking oil on a direct injection engine. The decrement in CO emissions was noted at lower loads and increment at higher loads. Diesel depicted peak CO emissions, whereas B30 depicted the lowest CO emissions. Diminishing CO emissions of the blends were attributed to their high oxygen content in the biodiesels as well as less carbon in the blends. Oxygen assists in better combustion of fuel resulting in reduced CO emissions. Similar results were reported for used waste cottonseed oil methyl esters by (Sinha et al., 2016).CO emissions increased exponentially at higher loads. Diesel showed the highest CO emissions, whereas the B20 showed the least CO emissions. A total of six fuels were tested, including neat diesel at varying speeds of the engine (Usta et al., 2004). Blend with 17.5%of biodiesel was reported to give maximum power. So the author chose to show the results of this blend against diesel for the sake of clarity. CO emissions for both the blend and the diesel decreased with the increase in the speed of the engine for all three loads. CO emissions were minimum at 50% load both for diesel and the blend, as their curves were very close to each other. At 75% load and 100% load, the CO emissions of the blend depicted a rise till 2000rpm. But after 2000 rpm, there was a drastic reduction in the CO emissions of the blend as well as diesel at both the loads. At lower speed, the atomization of the blend would have been difficult because of its high viscosity and so locally accumulated rich mixture accounts to incomplete combustion. At a higher speed, more air was supplied by the turbocharged engine, which increased turbulence. The air-fuel mixture was mixed uniformly, which resulted in complete combustion at a higher speed. (Al Dawody et al., 2014) prepared blends of soybean methyl esters (20%, 40%, and 100%) and tested against diesel, and results were compared with the R-K diesel software. Pure soybean biodiesel showed the least CO emissions, and diesel fuel reported the highest CO emissions.

Effect of Biodiesel on HC Emissions

Hydrocarbons are formed due to locally rich or locally lean mixture during the combustion. The locally lean mixture does not take part in propagating flame as they do not auto-ignite. These sites consume combustible mixture only if quenched by the boundary layer of rapid mixing air. The complete combustion of the locally rich mixture depends upon the availability of more air before the quick expansion (Chehroudi, B.,). Most of the kinds of literature biodiesel blends have reported diminished HC emissions. However, (Adaileh et al., 2012) claimed increment in HC emissions for all waste cooking oil biodiesel blends with rising loads, but emissions were still low against diesel. Increment in HC emissions at small engine loads and then declining HC emissions at higher loads for biodiesel were claimed by (Shirneshan et al., 2013) using waste frying oil biodiesel. HC emissions rise for biodiesel blend till 40% load, and then it continuously diminished for the rest of the loads.

Interestingly diesel depicted a decreasing trend at loads. However, biodiesel blends showed fewer emissions than diesel, and the B80 blend showed the least emissions. Oxygenated fuel of oxygen-rich air accounts to lower HC emissions in the case of biodiesel. Though biodiesel is less volatile than diesel, distillation points for diesel are higher, which prevents the final fraction from entirely vaporizing and burning, resulting in increased emissions.

Effect of Biodiesel on NO_x Emissions

NO_x emission pivots on the type of feedstock used as biodiesel. Highly saturated fuels emit fewer NO_x emissions. A reduction of 38.5% of NO_x was reported for B100 tallow methyl ester (Oner et al., 2009). (Wyatt et al., 2005) experimented with biodiesel blends derived from three animal fat and compared it with soy-based blends. The animal fat-based biodiesel B20 showed a reduction in NO_x levels than blends of soybean biodiesel. (Nabi et al., 2006) registered higher NO_x emissions with increased neem oil biodiesel proportions in diesel. Higher oxygen content accounted for an increase in NO_x emissions of biodiesel blends. EGR system is useful in curbing NO_x emissions with biodiesel as it controls oxygen content, which reduces the effective air-fuel ratio. Emissions of 15% neem biodiesel blend were found to be less than diesel with EGR. When fresh intake air to the engine is mixed with the exhaust gases, its specific heat increases, which reduces the combustion temperature (Yasina et al., 2017). Cetane number of fuel also affects NO_x emissions. Increased cetane number results in decreased ignition delay, which requires less fuel energy in the premixed stage.

Consequently lower peak cylinder gas temperature is produced, which controls combustion and hence fewer NO_x emissions (Muneeswaran et al., 2015). Advancing injection timing also increases NO_x emissions. (Senthil et al., 2015) tested Annona methyl ester blend at different injection timings of 24 ° bTDC, 27 ° bTDC, 30 ° bTDC, 33 ° bTDC. Both diesel and Annona methyl ester (A20) showed the least NO_x emission at 30°bTDC injection timing. However, NO_x emission of A20 was a little higher than diesel due to the high combustion temperature of the fuel. The load and speed of the engine also influence NO_x emissions (Murillo et al., 2007). As the load and speed increased, NO_x emissions of used cooking oil biodiesel samples and diesel decreased due to faster combustion at higher speed caused due to high turbulence inside the combustion chamber.

Effect of Biodiesel on Particulate Matter

Considerable logical proof exists on the negative impacts of diesel exhaust on human health. Researchers are worried about both malignant growth and noncancerous impacts from diesel exhaust. An ongoing epidemiological investigation of under the earth excavators discovered an expanded danger of alveolus malignant growth mortality related to diesel (Traviss et al., 2012). The main reason for soot emission is fuel containing 12 to 22 carbon atoms with hydrogen to carbon atom ratio as 2(Heywood,1988). Many data suggest that biodiesel can have a positive effect on the health of humans (Xiu et al., 2011; Swanson et al., 2007; McCormick et al., 2006). It has been reported that biodiesel reduces PM in the tail exhaust due to lack of aromatic hydrocarbons and Sulphur and also its high oxygen content. For conventional rail passenger cars, diesel engines with high-pressure injection fueled with B70 biodiesel; the particulate matter was found to be less than diesel (Tziourtzioumis et al.,2009). Biodiesel has shown significant PM reductions in heavy-duty vehicles. (Wang et al., 2000) used soyate methyl ester blend B35 against clean diesel in the unmodified truck engine. A reduction of 25% in PM emissions was reported. This reduction can be due to low Sulphur content and lack of aromatic content in the biodiesel.

CONCLUSION

Feedstock for biodiesel depends upon its availability in that region. Biodiesel is renewable and sustainable and is the future scope for the transport sector. Properties of biodiesel and also the type of engine used for test plays a vital role in performance and emissions. Added to it, factors like type and grade of diesel used, measurement techniques, type of instrument also influence the performance and emissions of biodiesel.

The following conclusions may be drawn from the above study.

1. High-density biodiesel has resulted in higher brake thermal efficiency. Loss of power and increase in fuel consumption is accounted to lower heating values of biodiesel. Hence biodiesel properties like heating value, density, viscosity, and cetane number play an essential role in performance characteristics.
2. The type of biodiesel significantly affects NO_x emissions. Unsaturated fuel, as well as a saturated ester with shorter chain length, produces more NO_x emissions. An increase in NO_x is also due to more oxygen content of the biodiesel, which increases the combustion temperature inside the cylinder. Advancing injection timing encourages the advanced start of combustion, leading to a high peak temperature and thereby producing more NO_x. Higher cetane causes decreased premix combustion leading to low temperature and pressure gradients which reduces the NO_x emissions
3. Decreasing trends in PM, HC, and CO emissions are noted with biodiesel as a fuel in CI engines. The additional oxygen content of biodiesel aids incomplete combustion and thereby reduced CO and HC emissions. Cetane number is also one of the critical factors in the reduction of CO and HC emissions. It is commonly accepted that CO emissions reduce when using biodiesel, due to its lower C: H ratio and high oxygen. Injection strategies and combustion timing also influences HC, CO and PM emissions
4. Early start of combustion (SOC) and shorter ignition delay is noted with biodiesel as compared to diesel due to its complex and rapid preflame reaction at higher temperatures. An increase in the maximum heat release rate was observed for biodiesel–diesel blend compared to diesel. The peak cylinder gas pressure is slightly higher than that of ordinary diesel fuel. It also depends upon the amount of fuel injected inside the cylinder.

REFERENCES

Abed, K. A., Akila K. El Morsi, Mahmoud M. Sayed, A. A. El Shaib, and M. S. Gad. "Effect of waste cooking-oil biodiesel on performance and exhaust emissions of a diesel engine." *Egyptian journal of petroleum* 27, no. 4 (2018): 985-989.

Adaileh, Wail M., and Khaled S. Al Qdah. "Performance of diesel engine fuelled by a biodiesel extracted from a waste cocking oil." *Energy Procedia* 18 (2012): 1317-1334.

Al Dawody, Mohamed F., and S. K. Bhatti. "Experimental and computational investigations for combustion, performance and emission parameters of a diesel engine fueled with soybean biodiesel-diesel blends." *Energy Procedia* 52 (2014): 421-430.

Arbab, M.I., Masjuki, H.H., Varman, M., Kalam, M.A., Imtenan, S. and Sajjad, H., Fuel properties, engine performance and emission characteristic of common biodiesels as a renewable and sustainable source of fuel. *Renewable and Sustainable Energy Reviews*, Vol.22, pp.133-147(2013).

Bittle, J. A., B. M. Knight, and T. J. Jacobs. "Interesting behavior of biodiesel ignition delay and combustion duration." *Energy & fuels* 24, no. 8 (2010): 4166-4177.

Bueno, André Valente, Mariana Paulinia Bento Pereira, João Victor de Oliveira Pontes, Francisco Murilo Tavares de Luna, and Célio Loureiro Cavalcante Jr. "Performance and emissions characteristics of castor oil biodiesel fuel blends." *Applied Thermal Engineering* 125 (2017): 559-566.

Çelikten, İsmet, Emre Mutlu, and Hamit Solmaz. "Variation of performance and emission characteristics of a diesel engine fueled with diesel, rapeseed oil and hazelnut oil methyl ester blends." *Renewable Energy* 48 (2012): 122-126.

Chang, J., Dennis Y. C. Leung, C. Z. Wu, and Z. H. Yuan. "A review on the energy production, consumption, and prospect of renewable energy in China." *Renewable and Sustainable Energy Reviews* 7, no. 5 (2003): 453-468.

Correa, Sergio Machado, and Graciela Arbilla. "Carbonyl emissions in diesel and biodiesel exhaust." *Atmospheric environment* 42, no. 4 (2008): 769-775.

Dwivedi, Gaurav, M. P. Sharma, and Mukesh Kumar. "Status and policy of biodiesel development in India." *International Journal of Renewable Energy Research (IJRER)* 4, no. 2 (2014): 246-254.

Gautam, Anirudh, and Avinash Kumar Agarwal. "Experimental investigations of comparative performance, emission and combustion characteristics of a cottonseed biodiesel-fueled four-stroke locomotive diesel engine." *International Journal of Engine Research* 14, no. 4 (2013): 354-372.

Gopal, K. Nantha, and R. Thundil Karupparaj. "Effect of pongamia biodiesel on emission and combustion characteristics of DI compression ignition engine." *Ain Shams Engineering Journal* 6, no. 1 (2015): 297-305.

Gumus, M. "A comprehensive experimental investigation of combustion and heat release characteristics of a biodiesel (hazelnut kernel oil methyl ester) fueled direct injection compression ignition engine." *Fuel* 89, no. 10 (2010): 2802-2814.

Knothe, Gerhard, Andrew C. Matheaus, and Thomas W. Ryan III. "Cetane numbers of branched and straight-chain fatty esters determined in an ignition quality tester." *Fuel* 82, no. 8 (2003): 971-975.

Kumar, Ashwani, and Satyawati Sharma. "Potential non-edible oil resources as biodiesel feedstock: an Indian perspective." *Renewable and Sustainable Energy Reviews* 15, no. 4 (2011): 1791-1800.

Mirza, Umar K., Nasir Ahmad, and Tariq Majeed. "An overview of biomass energy utilization in Pakistan." *Renewable and Sustainable Energy Reviews* 12, no. 7 (2008): 1988-1996.

Moka, Sudheshna, Maneesha Pande, Monika Rani, Ruchi Gakhar, Madhur Sharma, Jyoti Rani, and Ashok N. Bhaskarwar. "Alternative fuels: an overview of current trends and scope for future." *Renewable and Sustainable Energy Reviews* 32 (2014): 697-712.

Murillo, S., J. L. Miguez, J. Porteiro, E. Granada, and J. C. Moran. "Performance and exhaust emissions in the use of biodiesel in outboard diesel engines." *Fuel* 86, no. 12-13 (2007): 1765-1771.

Nabi, Md. Nurun, Md. Shamim Akhter, and Mhia Md. Zaglul Shahadat. "Improvement of engine emissions with conventional diesel fuel and diesel–biodiesel blends." *Bioresource Technology* 97, no. 3 (2006): 372-378.

Nair, Jayashri N., Ajay Kumar Kaviti, and Arun Kumar Daram. "Analysis of performance and emission on compression ignition engine fuelled with blends of neem biodiesel." *Egyptian Journal of Petroleum* 26, no. 4 (2017): 927-931.

Nayak, Swarup Kumar, and Bhabani Prasanna Pattanaik. "Experimental investigation on performance and emission characteristics of a diesel engine fuelled with mahua biodiesel using additive." *Energy Procedia* 54, no. 0 (2014): 569-79.

Ndayishimiye, Pascal, and Mohand Tazerout. "Use of palm oil-based biofuel in the internal combustion engines: performance and emissions characteristics." *Energy* 36, no. 3 (2011): 1790-1796.

Öner, Cengiz, and Şehmus Altun. "Biodiesel production from inedible animal tallow and an experimental investigation of its use as alternative fuel in a direct injection diesel engine." *Applied energy* 86, no. 10 (2009): 2114-2120.

Pandhare, Amar, and Atul Padalkar. "Investigations on performance and emission characteristics of diesel engine with biodiesel (Jatropha Oil) and its blends." *Journal of renewable Energy* 2013 (2013).

Panwar, N. L., Hemant Y. Shrirame, N. S. Rathore, Sudhakar Jindal, and A. K. Kurchania. "Performance evaluation of a diesel engine fueled with methyl ester of castor seed oil." *Applied Thermal Engineering* 30, no. 2-3 (2010): 245-249.

Ramadhas, A. S., C. Muraleedharan, and S. Jayaraj. "Performance and emission evaluation of a diesel engine fueled with methyl esters of rubber seed oil." *Renewable energy* 30, no. 12 (2005): 1789-1800.

Rao, G. Lakshmi Narayana, S. Sampath, and K. Rajagopal. "Experimental studies on the combustion and emission characteristics of a diesel engine fuelled with used cooking oil methyl ester and its diesel blends." *International Journal Engineering and Applied Sciences* 4, no. 1 (2008): 64-70.

Rao, Y. V. Hanumantha, Ram Sudheer Voleti, A. V. Sitarama Raju, and P. Nageswara Reddy. "Experimental investigations on jatropha biodiesel and additive in diesel engine." *Indian Journal of Science and Technology* 2, no. 4 (2009): 25-31.

Sahoo, P.K. and Das, L.M., 2009. Combustion analysis of Jatropha, Karanja and Polanga based biodiesel as fuel in a diesel engine. *Fuel*, *88*(6), pp.994-999.

Sánchez-Borrotoa, Y., R. Piloto-Rodrigueza, M. Errastia, R. Sierensb, and S. Verhelst. "2013 ISES Solar World Congress Prediction of cetane number and ignition delay of biodiesel using Artificial Neural Networks." *Energy Procedia* 57 (2014): 877-885.

Santos, Bjorn S., Sergio C. Capareda, and Jewel A. Capunitan. "Engine performance and exhaust emissions of peanut oil biodiesel." *Journal of Sustainable Bioenergy Systems* 3, no. 04 (2013): 272.

Senthil, Ramalingam, Rajendran Silambarasan, and Nattan Ravichandiran. "Influence of injection timing and compression ratio on performance, emission and combustion characteristics of Annona methyl ester operated diesel engine." *Alexandria engineering journal* 54, no. 3 (2015): 295-302.

Shahabuddin, M., A. M. Liaquat, H. H. Masjuki, M. A. Kalam, and M. Mofijur. "Ignition delay, combustion and emission characteristics of diesel engine fueled with biodiesel." *Renewable and Sustainable Energy Reviews* 21 (2013): 623-632.

Shehata, Mohamed Saied. "Emissions, performance and cylinder pressure of diesel engine fuelled by biodiesel fuel." *Fuel* 112 (2013): 513-522.

Shirneshan, Alireza. "HC, CO, CO2 and NOx emission evaluation of a diesel engine fueled with waste frying oil methyl ester." *Procedia-Social and Behavioral Sciences* 75 (2013): 292-297.

Sinha, Duple, and S. Murugavelh. "Biodiesel production from waste cotton seed oil using low cost catalyst: Engine performance and emission characteristics." *Perspectives in science* 8 (2016): 237-240.

Sinha, S., and A. K. Agarwal. "Experimental investigation of the combustion characteristics of a biodiesel (rice-bran oil methyl ester)-fuelled direct-injection transportation diesel engine." *Proceedings of the Institution of Mechanical Engineers, Part D: Journal of Automobile Engineering* 221, no. 8 (2007): 921-932.

Sivalakshmi, Sivanathan, and T. Balusamyb. "Experimental investigation on a diesel engine using neem oil and its methyl ester." *Thermal science* 15, no. 4 (2011): 1193-1204.

Swanson, Kimberly J., Michael C. Madden, and Andrew J. Ghio. "Biodiesel exhaust: the need for health effects research." *Environmental health perspectives* 115, no. 4 (2007): 496-499.

Tesfa, Belachew, Rakesh Mishra, C. Zhang, F. Gu, and A. D. Ball. "Combustion and performance characteristics of CI (compression ignition) engine running with biodiesel." *Energy* 51 (2013): 101-115.

Traviss, Nora. "Breathing easier? The known impacts of biodiesel on air quality." *Biofuels* 3, no. 3 (2012): 285.

Tziourtzioumis, Dimitris, Loucas Demetriades, Olympia Zogou, and A. M. Stamatelos. "Experimental investigation of the effect of a B70 biodiesel blend on a common-rail passenger car diesel engine." *Proceedings of the Institution of Mechanical Engineers, Part D: Journal of Automobile Engineering* 223, no. 5 (2009): 685-701.

Umesh, T., H. N. Manjunath, P. Rukmangadha, and D. Madhu. "Experimental study of performance and emission analysis of rice bran oil as an alternative fuel for an IC engine." *IOSR Journal of Mechanical and Civil engineering* 11 (2014): 130-134.

Usta, N., E. Öztürk, Ö. Can, E. S. Conkur, S. Nas, A. H. Con, A. Ç. Can, and M. Topcu. "Combustion of biodiesel fuel produced from hazelnut soapstock/waste sunflower oil mixture in a diesel engine." *Energy conversion and management* 46, no. 5 (2005): 741-755.

Valente, Osmano Souza, Vanya Márcia Duarte Pasa, Carlos Rodrigues Pereira Belchior, and José Ricardo Sodré. "Exhaust emissions from a diesel power generator fuelled by waste cooking oil biodiesel." *Science of the total environment* 431 (2012): 57-61.

Wang, Wei-Guang, Donald W. Lyons, Nigel N. Clark, M. Gautam, and P. M. Norton. "Emissions from nine heavy trucks fueled by diesel and biodiesel blend without engine modification." *Environmental science & technology* 34, no. 6 (2000): 933-939.

Wood, Benjamin M., Kerry Kirwan, Steven Maggs, James Meredith, and Stuart R. Coles. "Study of combustion performance of biodiesel for potential application in motorsport." *Journal of Cleaner Production* 93 (2015): 167-173.

Xue, Jinlin, Tony E. Grift, and Alan C. Hansen. "Effect of biodiesel on engine performances and emissions." *Renewable and Sustainable energy reviews* 15, no. 2 (2011): 1098-1116.

Yasin, Mohd Hafizil Mat, Rizalman Mamat, Ahmad Fitri Yusop, Daing Mohamad Nafiz Daing Idris, Talal Yusaf, Muhammad Rasul, and Gholamhassan Najafi. "Study of a diesel engine performance with exhaust gas recirculation (EGR) system fuelled with palm biodiesel." *Energy Procedia* 110 (2017): 26-31.

Books

Heywood, J.B. *Internal Combustion Engine Fundamentals;* Volume 21; McGraw-Hill Education: New York, NY, USA, (1988)

McCormick, Robert L., Aaron Williams, John Ireland, and R. R. Hayes. *Effects of biodiesel blends on vehicle emissions: fiscal year 2006 annual operating plan milestone 10.4*. No. NREL/MP-540-40554. National Renewable Energy Lab. (NREL), Golden, CO (United States), 2006.

Website

Chehroudi, B., *Diesel Engine Emissions: Hydrocarbons(HC)* advtechconsultants.com.

In: Applied Soft Computing Techniques … ISBN: 978-1-53618-180-7
Editors: Amit Kumar Thakur et al.

Chapter 9

SUSTAINABLE DEVELOPMENT THROUGH LEARNING CLIMATE CHANGE: ADAPTATION MITIGATION AND FINANCE

Mahender Singh Kaswan**, Rajeev Rathi and Varun Panwar
School of Mechanical Engineering,
Lovely Professional University, Phagwara, Punjab, India

ABSTRACT

The increased level of greenhouse gases and an increased number of hazards in recent years has forced human society to rethink climate change, its effects, and current policy measures. There is an immense need to make aware of the community about climate change, mitigation, and adoption standards. This chapter provides a systematic way to understand climate change, adoption and mitigation measures, and climate finance. Moreover, the present work also describes the actions taken at the individual, organizational, societal, governmental level to mitigate climate emissions. This chapter will facilitate all the stakeholders

* Corresponding Author's Email: mahi622@rediffmail.com.

of environmental protection through a better understanding of various aspects of environments and various associated actions at different levels.

Keywords: climate change, environment, mitigation, adoption, climate finance, sustainability

1. Introduction

The increased level of emissions, untoward incidents, and health hazards are the result of severe climate change (World Metrological Organization, 2018). Climate tells us about what clothes we have to buy, whereas weather states what we have to wear (Rippke et al., 2016; Baleta et al., 2019). It is the result of the delicate energy balance. The temperature of our planet results from the balance between the energy coming into the earth from the sun and energy leaving outer space (Rodriguez et al., 2018; Nardone et al., 2010; Salvia et al., 2019). The current adopted anthropogenic activities, and with current policy measures, the temperature of the planet will rise by 3°Ctill the end of the century (Ramanathans et al., 2008; Durate et al., 2013). The consequence of environmental emission rise will be in more hazard, more gender equality, unanticipated events like storms and heavy rainfall, and deficient rain in different regions of the world. Therefore, society will be at the juncture of being adversely affected by the climate changed induced by the human being itself.

The rapid industrialization, reliance on non-renewable sources of energy, and other anthropogenic activities have increased the climate forcing by 41% from 1990 to 2017 (World Metrological Organization, 2018). Climate forcing is the factor that affects the temperature of the earth.

1.1. Main Drivers Affect the Temperature of the Earth

Three drivers affect the temperature on the surface of the planet are:

Solar variation-The small changes in the earth's rotation in orbit may change the amount of energy received by the Sun. However, in the past 140 years, we have experienced minimal solar variability.

Volcanic activities-The unpredicted volcanic eruptions can also cause climate disruptions. Significant eruptions inject aerosols into the atmosphere, which cools the earth's surface. Mt. Pinatubo volcanic eruption in 1991 has cooled the earth's surface by 0.1°C to 0.3°C for up to 3 years.

Carbon cycle-The increased emissions have a direct causal effect on change in the climate. Therefore, the increased emissions due to anthropogenic activities have led to substantially increased in the temperature of the earth in the last few centuries.

- CO_2 concentration

  The first decade of the 21st century was the warmest recorded since modern measurement began around 1850 (Tao et al., 2006; Stouffer et al., 2006). The increased level of CO_2 in the atmosphere, absorbed more reflected light from the earth's surface and increased the radiation absorbed by the planet. The level of CO_2 in 2016 was 403 parts per million (PPM), which was the highest among all the years of humankind (Wu et al., 2005).

1.2. Effect of Temperature Rise on Various Parameters

1.2.1. Sea Level Rise

Plate tectonics, ice caps on land melting or growing, ocean water warming/expanding, and cooling/contracting can affect the sea-level by changing the volume of ocean basins. The ocean capture 94% of the heat captured due to an increase in CO_2 level. In the past decade, the ocean absorbed around 25% of anthropogenic CO_2 emissions (Fletcher et al., 2010).

1.2.2. Ice Extent

Mountains are home to about 800 million people and are crucial for the provision of water and ecosystem services. Along with their altitude, mountains create different microclimates. On a bigger scale, they can also affect the local climate by acting as a climatic barrier to air currents or the monsoon flow (Zidansek et al., 2009). The most massive bodies of ice, the Greenland and Antarctic ice sheets, have been losing a large amount of ice at an increasing rate since 1992. Temperature increases are more substantial over land than over the ocean. Temperature increases are more significant at high altitudes than in the tropics.

1.2.3. Ocean Acidity

The ocean water absorbs the emitted CO_2 and leads to the acidification of the water body. Ocean surface pH has declined from 8.2 to 8.1 over the industrial due to the increased level of GHGs (Xu et al., 2008; Bardaji et al., 2009). This decline in the pH corresponds to a 30% increase in ocean acidity.

The chapter has organized into eight sections, including the introduction. The 2nddepicts the exposure and risk associated with climate change. Subsequently, the 3rd section demonstrates the adaptation to climate change. The mitigation measures have described in the 4th section of the chapter. The 5th section represents climate finance, and the 6th section depicts the contribution towards Mother Nature. The significant agreements and conventions related to climate change have presented in the 7th section of the manuscript. The last part of the chapter enumerated the conclusion.

2. Exposure and Risk Associated with Climate Change

The risk associated with climate change differs, it depends on the type of hazard, exposure to the danger, and vulnerability of the population

(Rind et al., 2005). The impact of the climate differs from place to place and from gender to gender (Zhang et al., 2008; Yuvnev et al., 2007). The hazard is a natural or human-induced physical event that may cause damage to the life, property, ecosystem, and other assets. The exposure is the presence of economic, social, natural, or cultural assets in places and settings that could be adversely affected by the hazard. The vulnerability is the predisposition of people or assets to be adversely affected or their capacity to cope and adapt (Zazo et al., 2008; Yang et al., 2011). Figure 1 depicts the risk associated with climate change in terms of measures.

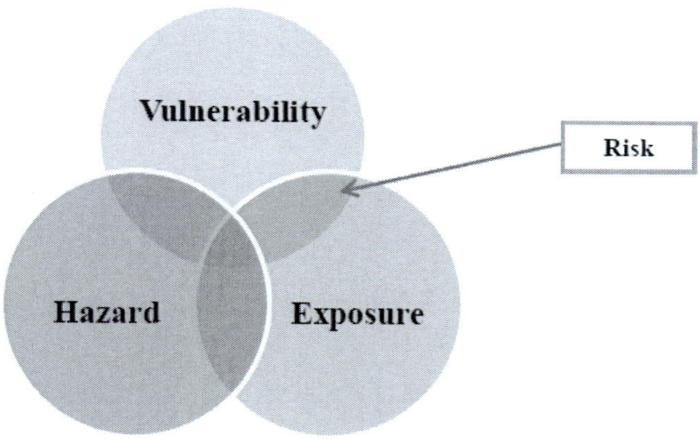

Figure 1. Cause of the impacts

2.1. Assess the Impact of Climate Change

Through the addition of the carbon data into climate models, the environmentalist comes up with different possible emission pathways. Before considering the impact of climate, it is a prerequisite to measure the predicted emission in the future. Emission pathways are trajectories of global anthropogenic emission over the 21st century (Intergovernmental Panel on Climate Change, 2014). Figure 2 depicts the CO_2 emission year wise from 1990 to 2016.

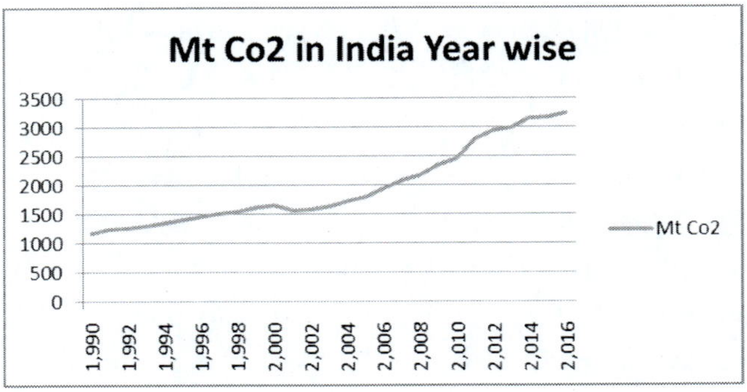

Figure 2. Projected CO_2 emission of India (climatewatchdata.org).

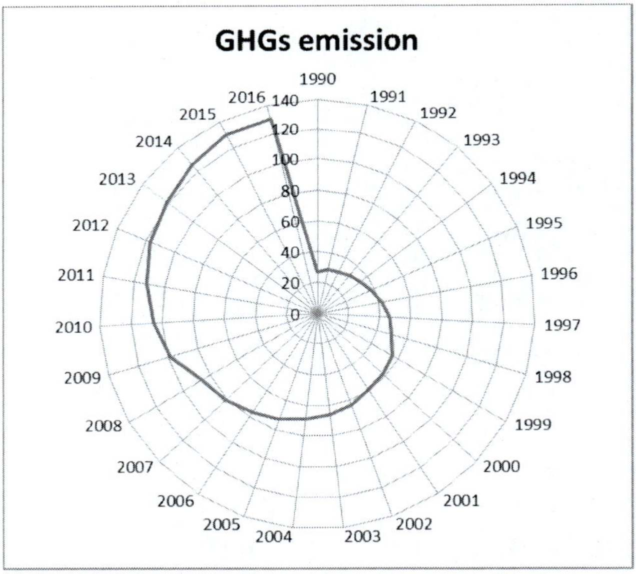

Figure 3. Radar chart of GHGs emissions in India from the industrial process.

The different global mean surface temperature changes are projected based on the amount of GHGs emissions. The four emission pathways, except no-climate-policy baseline, limit the temperature increase to 1.5°C to 2°C by 2100 (Bertrand et al., 2011; Govindan et al., 2020). Figure 3 demonstrates GHGs emissions from the industrial process for India. The discharge of GHGs had increased by nearly 500% from 1990.

3. Adaptation to Climate Change

The adaption to climate change is to adjust to the changing environment (IPCC, Special Report 1.5°C). Every adaptation process, whether planning for actions at the national level or designing a local adaptation project, is more sustainable when it made on a consultative process (Trenberth et al., 2018; Ahmed et al., 2019). In other words, effective adaptation action brings together in a meaningful way all the relevant stakeholders, beyond the operational team (Gu et al., 2012; Araos et al., 2016).

3.1. Types of adaptation approaches

3.1.1. Planned adaption or risk mitigation

This approach is applied when the climate risk is of high frequency and low or medium loss (Lobell et al., 2008; Aboulnaga and Mostafa, 2020). To cope with climate risk, Tulavu developed its National Adaptation Programs of Action.

3.1.2. Contingency adaptation or risk transfer

This approach is applied when the climate risk is of low frequency and medium to high loss. Some extreme events can fall into this category, such as long-term droughts (Berrang et al., 2019). Risk pooling mechanisms or approaches to manage migration flows might address this risk.

3.1.3. Loss acceptance or coping

This approach is applied when the hazard is devastating but very unlikely to happen. Severe extreme events can fall into this category, such as unprecedented cyclones (Ali and Erenstein, 2019).In these cases, relief support and humanitarian measures considered for a better cure.

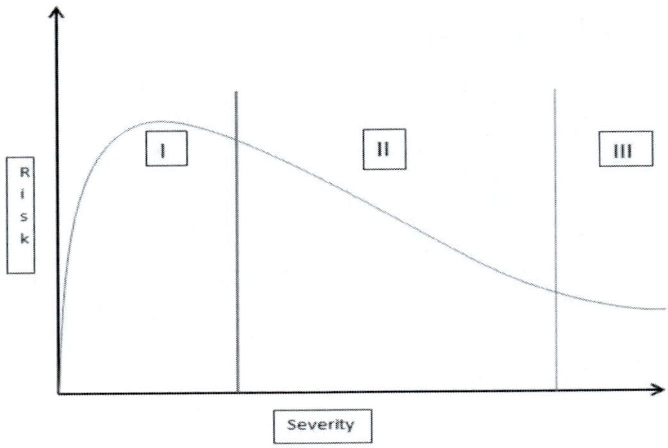

Figure 4. Adaptation approaches.

Figure 4 demonstrates the frequency and severity of the harm. In category 1, a high rate and low losses came, i.e., seasonal temperature change. The II category depicts low frequency and medium to high loss, i.e., the collapse of rain in the cropping season. The III category represents low frequency and high loss, i.e., low frequency and high loss, i.e., ultimate dessert for a more extended period.

4. MITIGATION

The mitigation means reducing the emission and increase the carbon sink (Demski et al., 2017). Sufficient work must initiate towards climate neutrality that means whatever the discharge we produce, the earth should have the capacity to absorb the same. Many international agencies and intergovernmental organizations are focusing on low carbon development policies that are compatible with the Paris agreement (Gill et al., 2018; Kaswan and Rathi, 2019). The small carbon development needs for technological, infrastructure, societal, and economic changeover (Mi et al., 2019). The carbon budget specifies the amount of maximum emission to meet The Paris agreement goal. The estimate carbon budget is 840 GT of

CO_2, and till now, 531 GT of CO_2 has already been emitted into the environment (Hasekamp et al., 2019). Therefore, to remain in the carbon budget, there is a need for solar technologies, low carbon investment, and measurement and tracking of GHGs. Until the 18th century, the atmospheric concentration of GHGs was found stable at 270 PPM. Since the industrial revolution, its level has been increasing enormously (Shan et al., 2018). In 2013, the concentration of the CO_2 was 400 PPM, and to meet the target of the Paris agreement, it should not go beyond 430 PPM till the end of this century (Ireland and Clausen, 2019). But it had been estimated that till 2040, CO_2 level projected to reach 450 PPM if we are continuing with current policy measures and technologies.

Two characteristics of gases determine the strength of their GHG effect are:

a. GWP

GWP is a measure of the radiative effect of a single unit of gas over a specified period. Table 1 depicts the GWP of different gases.

Table 1. GWP of the greenhouse gases

S. No.	Greenhouse gases	GWP
1	CH_4	28
2	N_2O	265
3	SF_6	23500

b. Atmospheric lifetime

The atmospheric lifetime is how long the gas stays in the atmosphere before the natural process removes it. A gas with a long lifetime can exert more warming influence than with a gas with a short lifetime (assuming the same GWP).

The energy sector has the most significant carbon footprint, with 32% of GHGs. The building consumes 40% of the primary energy and in building out of 40%, 70% by electricity. The mitigation for agriculture demands change in the agriculture practices, growth of local and adaptable crops, and avoid deforestation. The transport sector contributes 17% of the GHGs of the total produced (Wang et al., 2011; Feng et al., 2012). Traditionally, over the economy have been based on a linear model, used where raw materials extracted, products made, used. This economic model is inefficient, wasteful, and emission-intensive. Instead, we can design over economy around a circular model, where resources are renewable rather than finite, products and material are reused and recycled, waste from one industry becomes an input for the other, and pollution and emission ultimately eliminated.

4.1. Industrial Organizations Contribution towards Mitigation of the Emission

4.1.1. Estimate GHGs Emission

Carbon disclosure protocol helps over 2400 companies report on their emissions, and action was taken to reduce them.

4.1.2. Reduce GHGs Emissions

The companies can reduce the emission of the GHGs through

- Investment in low carbon technologies
- Switch to renewables
- Efficient appliances
- Support public transportation.

4.1.3. Offsetting/Compensation of Emissions

Industries can invest in projects that reduce emissions. The offsetting is just similar to the waste disposal.

4.1.4. Set a Price on Carbon

The industries must identify the inefficiency in big companies and help to address the same.

4.1.5. Divert from Fossil Fuels

The investor must invest in renewable-based energy-based projects.

5. CLIMATE FINANCE

According to the Global Commission on Adaptation, current ambition is not sufficient to ensure critical economic systems can cope with climate risks. Action to accelerate understanding, planning, and finance for climate resilience is needed, and the period before 2020 will be critical(Hong et al., 2020). The economy generates a considerable amount of resources near about 85 trillion USD. There is an SDG-Gap of 2.5 trillion/year that should spend on assessing the target of the Paris pact (Steckel et al., 2017). The global climate finance grew to almost 500 billion USD in 2018 due to the shift in the market supported by the government (Petersons and Skovgaard, 2019). Globally, much more is invested in mitigation than adaptation actions. Most of the climate investment has generated and spent with in the same country. Figure 5 depicts the year wise global climate finance in Bn USD. It has observed the global climate finance increased to 78.95% in 2018 as compared to 2013. Developed countries pledged to mobilize USD 100 billion per year by 2020, and through to 2025, to support developing countries. However, a recent report shows that flows from developed to developing country Parties increased by 14% in 2016 to USD 38 billion, which still falls short of the USD 100 billion *per year* target.

Figure 6 depicts the climate finance source and destination for the year 2015-2016. Climate change brings both risks and opportunities to businesses and financial institutions. It has deduced from figure six that most of the climate finance generated within the source country and also consumed in the same. So, there is an immense need for developed nations to spend more in developing countries for mitigation and adaptation measures.

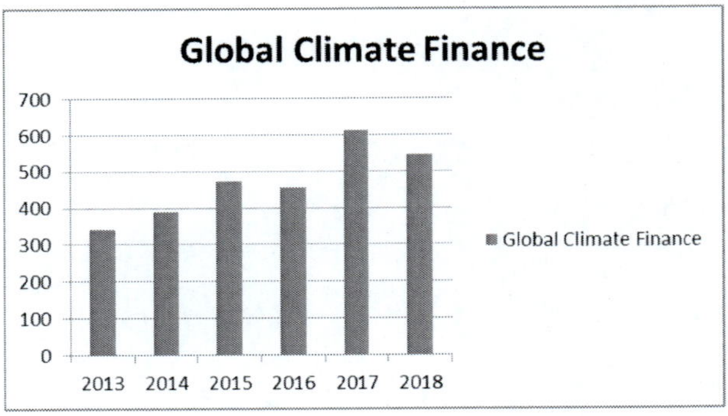

Figure 5. Year-wise global climate finance (in Bn USD).

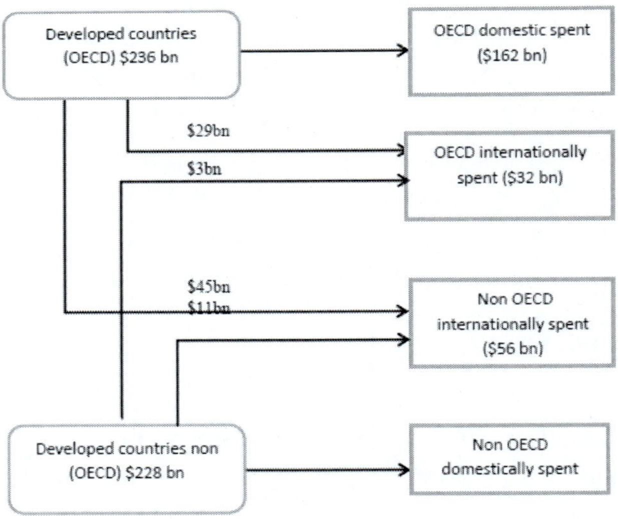

Figure 6. Climate Finance source and destination 2015-2016.

5.1. The Risk Associated with Climate Finance

5.1.1. Market and Policy Risks

Revenue streams and business operations are vulnerable to climate risks (Winkler and Dubash, 2016). This economic vulnerability depends on the type of industry and place of service. However, nowadays, it is felt by most market actors. Some main risks are as follows:

- Introduction of new climate policies that affect carbon price.
- Failure to mitigate climate impacts or disclose climate information,
- Changes in market price and availability of carbon-intensive goods and services may plummet with decreased demand.

5.1.2. Physical Risks

Damage to physical assets, natural capital, or people by climate change that can be:

- Generated by extreme events such as cyclones or floods.
- It has driven by long-term climate effects such as sea-level rise.

In the face of those risks, there is still a big insurance gap. Financial institutions are starting to recognize climate risks as financial risks but are still lagging on insuring against them (Jobst et al., 2019). The risks associated with an increase of temperature of over 2 degrees are significantly higher compared to 1.5 degrees, which interferes with the ability of businesses to plan and prepare. (Zhivkova, 2019; Kaswan and Rathi, 2020).

5.2. Opportunities Associated with Climate Finance

According to the Global Commission on the Economy and Climate, climate actions could bring at least 26 trillion USD in economic benefits by 2030. To harvest these benefits, the Commission advises on focusing on capturing the opportunities in critical sectors and embracing innovation.

5.2.1. The Sharing Economy

The sharing economy is the concept where the mutually sharing of the resources done to mitigate climate risk or to utilize the fund in the most effective ways. The resources are divided based on either free trial or on a reduced basis.

5.2.2. The Circular Economy

The circular economy is a modern concept to ensure environmental sustainability in the system through reduces, reuse and recycle of the products after their primary use by the customer (Bocken et al., 2016; Geissdoerfer et al., 2017). The circular economy concepts will enhance the organization's competitiveness, reduce the intergovernmental pressure on the organization to combat climate change, will enhance the resource efficacy, and in the run, will increase the profitability of the organizations (Lieder and Rashid, 2016; Ormazabal et al., 2020). Figure 6 depicts the concept of the circular economy.

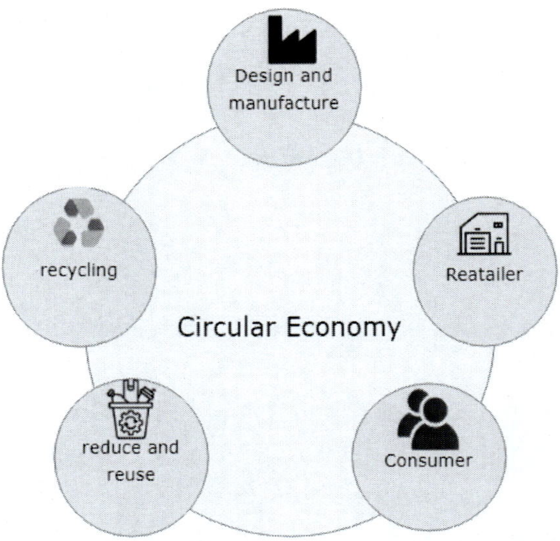

Figure 6. Circular economy.

i) Building resilient and inclusive societies

Climate resilience can yield economic, social, and environmental benefits. Here are a few examples:

- Sustainable urban planning, make living units more efficient and durable, while new mobility models both reduce climate impacts and raise citizens' well-being.
- Improving access to sustainable energy, water, and waste management systems creates jobs and business opportunities, decrease social tension, and enhance climate readiness and resilience.
- A sustainable food and land system (including forests) provides jobs to the most vulnerable, increases food security, decreases migration, and is a cost-effective way to capture and store carbon.

ii) Sustainable development goals

The Paris Agreement sets out the most ambitious plan for sustainable development. In this agreement, the parties agreed to set nationally determined contributions to mitigate climate challenges. The parties also agreed on the 17 sustainable development goals (SDGs) that line out the most determined plan for humanity to address climate issues.

6. Contribution to the Mother Nature

All levels of governance, as well as institutions and individuals, can contribute to creating the enabling conditions for climate policy to be effective. Everyone in society has a role to play! To begin with, both companies and individuals can calculate their carbon footprint. Secondly, both companies and individuals can reduce emissions, for example, by increasing energy efficiency or promoting climate-friendly transportation.

When this is not possible, both companies and individuals can offset their emissions. Both businesses and individuals can send necessary price signals by choosing to invest and spend their finances in a climate-friendly manner (Kaswan et al., 2019; Ferreira et al., 2020). Lastly, both businesses and individuals can raise awareness on climate action, and demand and support higher ambitions by governments and others in society. All levels of governance, as well as institutions and individuals, can contribute to creating the enabling conditions for climate policies to be effective.

6.1. Contribution from Individual

Individual action is needed to adopt sustainable lifestyles how we can all contribute to redefining economic behavior and reducing the carbon footprint include.

6.1.1. Switching or Reducing Home Energy Use

The individual has to shift towards more efficient fuels and renewable sources for the production of foods. It also demands efficient building design and move towards sustainable and effective heat transfer devices.

6.1.2. Hanging Diet

The climate change mitigation demands from an individual household in terms of better eating habits. The energy used for the baking of one kg of meat is 19 times more than the making of the same quantity of cereals. Therefore, to contribute to Mother Nature, human beings have to shift towards vegetarian foods.

6.1.3. Rethinking Travel and Mobility

Climate change pursuits demand the effective utilization of transports and mobility. The government is investing more in renewable operated transportation measures. It should be the moral duty of every human being that they should use public transport instead of the private one.

6.1.4. Buying Less and Reusing More

The reusing culture should be adopted in daily life as most of the products after primary use is either buried or burned that leads to the emission of harmful GHGs in the environment.

6.1.5. Educating Family and Friends

It is the primary duty of everyone to protect the coming nature through educating the people on sustainable measures, ecofriendly products, alternative methods of production of energy, and methods.

6.2. Contribution from Institutions

Institutions and companies can go a long way to promote climate action in their business models and operations (Landholm et al., 2019). On the one hand, companies are at the front end of innovation and finding sustainable alternatives for current energy, tools, and transport models. But, they can also set an example within their operations by greening the supply chain, implementing green procurement, and promoting recycling.

6.3. Actions by Regions and States

In countries with a robust federal or state system such as in the USA, regions have budget and policy freedom that allows them to formulate and implement climate actions. Analysis by the carbon disclosure project shows that stated and areas are on the average, setting a more ambitious target than those set by the national counterparts.

6.4. Actions by Countries

The government of each country is responsible for setting up policies, strategies, and measures that guide economic development while ensuring environmental and social protection. In recent years, most countries shave factored climate change into their development policies. However, political instability, social and economic unrest have thawed climate ambitions. Policy action national governments are not the only political actors that can initiate climate action. There are many positive examples of climate action led by state, regional, and city authorities. However, governments are a key player in climate action architecture. Through clear political vision and leadership, national governments can drive consensus at an international level, as well as translate climate urgency into local and regional action.

There is a myriad of policy tools that governments can employ to foster climate change mitigation. As Miguel explained, policy tools can group into several categories.

6.4.1. Regulatory Tools

Governments can establish bans or limits, and introduce standards on products or processes to make production and consumption patterns more sustainable. Regulatory tools may include:

- Environmental standards
- Technical/emissions standards
- Restrictions and bans
- Property laws and access rights

Most countries, for example, have had emission standards for vehicles for many years. However, a growing number of states are announcing a ban on the sale of fossil fuel-powered cars (combustion engine), sending a clear signal to consumers and producers.

6.4.2. Economic and Market Tools

These tools aim to correct market failures by factoring the environmental and social costs into the pricing of goods and services and incentivizing/rewarding green businesses and sustainable practices. Table 2 depicts the economic and market tools adopted by the government to take steps towards sustainable development.

Table 2. Economic and market tools

Economic incentives ("carrots")	Full Cost Pricing ("sticks")
Tax rebates and exemptions	Environmental taxes
Direct financial support (subsidies, grants, etc.)	Polluters' pay charges
Public investment (public projects, green bonds, loans, feed-in tariffs, etc.)	Users' pay charges
Sustainable public procurement	Removal of harmful subsidies
	Carbon trading

In recent years, the size of fossil fuel global subsidies has been decreasing. However, it is still more than twice that of grants available for renewable energy. (International Energy Agency, 2019).

6.4.3. Information and Awareness-Raising Tools

These instruments provide knowledge on the environmental performance of products, services, or systems. They also help both consumers and producers to improve their understanding of climate change challenges and the government's climate policy and formulate climate action accordingly. Examples of such systems include:

- Eco-labeling
- Sustainability reporting
- Customer advice centers
- Awareness-raising campaigns
- Training and education for climate action

6.4.4. Voluntary Action

- Voluntary agreements imply that firms make commitments going beyond legal requirements. Voluntary agreements can be:
- Sectorial voluntary agreements
- Negotiated agreements (i.e., with the participation of public institutions)
- Unitary voluntary agreement

6.4.5. Action at the International Level

The Paris agreement makes responsible countries in their fight for climate change. It is the first-ever legally bind binding global climate deal adopted by 195 countries. The primary financial mechanism to implement the Paris agreement is the green climate fund. The nations have also set up several blocks, contracts, and partnerships to coordinate and advance climate change action.

7. Major Agreements

7.1. First World Climate Conference

The first pursuits to address the mitigation and adaptation knowledge on climate can be traced back to 1979. An increase in scientific evidence and growing public concern pushed the matter of human being interaction with Mother Nature into the political agenda in the mid-1980s. Governments concluded that there was a need for an impartial and independent body to address this issue, leading to the creation of the Intergovernmental Panel on Climate Change (IPCC) in 1988.

7.2. UN General Assembly (UNGA)

The UNGA makes rigorous pursuits and vows to mitigate climate change through the knowledge share, innovation, and recycling of the materials. The IPCC was mandated to compile a comprehensive review of the state of knowledge of climate change science and to provide recommendations of the social and economic impacts of climate change. The IPCC was also mandated to identify possible response strategies.

7.3. IPCC First Assessment Report

The UNGA formally launched negotiations on a framework convention that was conducted by an intergovernmental negotiating committee (INC), which met for the first time in 1991. Fifteen months later, the INC adopted the UNFCCC by consensus.

7.4. UN Framework Convention

The UNFCCC was adopted in 1992 to address the global climate issues, to bring the entire world on a single platform to address the climate issues, and develop mitigation and adaptation measures for the well-being of everyone in the society.

7.4.1. Kyoto Protocol

The Kyoto Protocol is an international agreement under the Convention that was adopted in 1997 and entered into force in 2005 (Quershi et al., 2019). It committed its Parties (mostly developed countries) to set internationally binding emission reduction targets. The Doha amendment to the protocol passed in 2012 and the Parties entered a second commitment period of emission reductions. The Kyoto Protocol allowed countries to meet reduction targets through market-based mechanisms such as international emissions trading and clean development mechanisms for offsetting emissions. The Protocol includes a registry system to track and record transactions by Parties under the tools. In 2015, the Kyoto Protocol was succeeded by the Paris agreement (see the next section of the course).

7.4.2. Paris Pact

The COP 21 adopted the Paris pact to achieve a world that is carbon neutral and resilient to the climate change by mid-century and keep the increase in global mean temperature to well below 2°C. The Paris pact is unanimous agreement among the world leaders that every country, apart from their own needs to do much more to save the planet. The Paris pact has a total of 197 parties making pursuits for climate care. The private parties have set up nationally determined contributions (NDCs) to mitigate the climate issues. NDCs are climate plans every single country is doing on its responsibility, defining and implementing as national agenda. Countries' NDCs are legally binding international instruments. Once in a given period, NDC or climate plans need to be renewed. The mechanism established under the Paris pact only knows one direction, i.e., upwards in terms of ambition. So, NDCs or climate plans have put in a long term perspective, and that will create pathways needed for sustainable development. The NDCs are different for developed and developing countries. For the developing nations, those GDP high dependent on the agriculture sector, then land use, reforestation, and restoration of soil might be the critical points of the plan. Meanwhile, for the developing countries, the emission reduction from the highest emitting sector must be the utmost priority. The global community, once a while, make the global stocktake or reality check for the achievement of NDCs. The first global stocktake will take place in 2023, and after each five, it will take place after every five years.

CONCLUSION

Since the last few decades, emission of GHGs emissions, untoward climate risks, and demands for sustainable products has increased substantially. Sustainable development and fighting climate change are not a trade-off but two sides of the same coin. The industrial organizations, community as well as individual, have to contribute in the most effective way to achieve the ambitious target of the Paris Pact. There is a need for knowledge sharing on climate change, more investment in adaptation, sustainable lifestyles, and eating habits. The industrial organizations have to device alternative sources of energy, less carbon footprint production methods, and recycling technologies. The intergovernmental organizations have to set more rigorous policy measures than ever before to limit the global surface temperature up to 2°C.

REFERENCES

Aboulnaga, M., & Mostafa, M. (2020). Climate Change Adaptation: Prioritising Districts for Urban Green Coverage to Mitigate High Temperatures and UHIE in Developing Countries. In *Renewable Energy and Sustainable Buildings* (pp. 825-837). Springer, Cham.

Ahmed, A. U., Appadurai, A. N., & Neelormi, S. (2019). Status of climate change adaptation in the South Asia region. In *Status of climate change adaptation in Asia and the Pacific* (pp. 125-152).Springer, Cham.

Ali, A., & Erenstein, O. (2017).Assessing farmer use of climate change adaptation practices and impacts on food security and poverty in Pakistan. *Climate Risk Management*, *16*, 183-194.

Araos, M., Berrang-Ford, L., Ford, J. D., Austin, S. E., Biesbroek, R., & Lesnikowski, A. (2016). Climate change adaptation planning in large cities: a systematic global assessment. *Environmental Science & Policy*, *66*, 375-382.

Baleta, J., Mikulčić, H., Klemeš, J. J., Urbaniec, K., & Duić, N. (2019).Integration of energy, water, and environmental systems for sustainable development. *Journal of cleaner production*, *215*, 1424-1436.

Bardají, T., Goy, J. L., Zazo, C., Hillaire-Marcel, C., Dabrio, C. J., Cabero, A., & Lario, J. (2009). Sea level and climate change during OIS 5e in the Western Mediterranean. *Geomorphology*, *104*(1-2), 22-37.

Berrang-Ford, L., Biesbroek, R., Ford, J. D., Lesnikowski, A., Tanabe, A., Wang, F. M.,&Grecequet, M. (2019). Tracking global climate change adaptation among governments. *Nature Climate Change*, *9*(6), 440-449.

Bertrand, R., Lenoir, J., Piedallu, C., Riofrío-Dillon, G., de Ruffray, P., Vidal, C., & Gégout, J. C. (2011). Changes in plant community composition lag behind climate warming in lowland forests. *Nature*, *479*(7374), 517-520.

Bocken, N. M., De Pauw, I., Bakker, C., & van der Grinten, B. (2016). Product design and business model strategies for a circular economy. *Journal of Industrial and Production Engineering*, *33*(5), 308-320.

Change I. C. (2014).Mitigation of climate change. *Contribution of Working Group III to the Fifth Assessment Report of the Intergovernmental Panel on Climate Change*, *1454*.

Demski, C., Capstick, S., Pidgeon, N., Sposato, R. G., & Spence, A. (2017). Experience of extreme weather affects climate change mitigation and adaptation responses. *Climatic Change*, *140*(2), 149-164.

Duarte, C. M., Losada, I. J., Hendriks, I. E., Mazarrasa, I., & Marbà, N. (2013).The role of coastal plant communities for climate change mitigation and adaptation. *Nature Climate Change*, *3*(11), 961-968.

Energy, G. (2019). *CO2 status Report*. IEA (International Energy Agency): Paris, France.

Feng, S., Ho, C. H., Hu, Q., Oglesby, R. J., Jeong, S. J., & Kim, B. M. (2012). Evaluating observed and projected future climate changes for the Arctic using the Köppen-Trewartha climate classification. *Climate Dynamics*, *38*(7-8), 1359-1373.

Fletcher, W. J., Goñi, M. S., Peyron, O., & Dormoy, I. (2010*). Abrupt climate changes of the last deglaciation detected in a Western Mediterranean forest record.*

Geissdoerfer, M., Savaget, P., Bocken, N. M., & Hultink, E. J. (2017). The Circular Economy–A new sustainability paradigm? *Journal of cleaner production*, *143*, 757-768.

Gil, J. D. B., Daioglou, V., van Ittersum, M. K., Reidsma, P., & van Vuuren, D. (2018, April). Reconciling global sustainability targets and regional action for food security and climate change mitigation. In the *7th AgMIP Global Workshop*.

Govindan, K., Shankar, K. M., & Kannan, D. (2020).Achieving sustainable development goals through identifying and analyzing barriers to the industrial sharing economy: A framework development. *International Journal of Production Economics*, *227*, 107575.

Gu, H., Wang, G., Yu, Z., & Mei, R. (2012). Assessing future climate changes and extreme indicators in east and south Asia using the RegCM4 regional climate model. *Climatic Change*, *114*(2), 301-317.

Hasekamp, O. P., Gryspeerdt, E., & Quaas, J. (2019). Analysis of polarimetric satellite measurements suggests stronger cooling due to aerosol-cloud interactions. *Nature communications*, *10*(1), 1-7. https://climatepolicyinitiative.org/publication/global-climate-finance-an-updated-view-2018/.

IPCC, 2014: *Climate Change 2014: Mitigation of Climate Change. Contribution of Working Group III to the Fifth Assessment Report of the Intergovernmental Panel on Climate Change*.

Ireland, P., & Clausen, D. (2019). Local action that changes the world: fresh perspectives on climate change mitigation and adaptation from Australia. In *Managing Global Warming* (pp. 769-782). Academic Press.

Kaswan, M. S., & Rathi, R. (2019). Analysis and modeling the enablers of green lean six sigma implementation using interpretive structural modeling. *Journal of cleaner production*, *231*, 1182-1191.

Kaswan, M. S., Rathi, R., & Singh, M. (2019). Just in time elements extraction and prioritization for health care unit using the decision-making approach. *International Journal of Quality & Reliability Management*.

Kaswan, M. S., & Rathi, R. (2020). Investigating the enablers associated with the implementation of Green Lean Six Sigma in the manufacturing sector using Best Worst Method. *Clean Technologies and Environmental Policy*, 1-12.

Landholm, D. M., Holsten, A., Martellozzo, F., Reusser, D. E., & Kropp, J. P. (2019). Climate change mitigation potential of community-based initiatives in Europe. *Regional Environmental Change*, *19*(4), 927-938.

Lieder, M., & Rashid, A. (2016). Towards circular economy implementation: a comprehensive review in the context of the manufacturing industry. *Journal of cleaner production*, *115*, 36-51.

Lobell, D. B., Burke, M. B., Tebaldi, C., Mastrandrea, M. D., Falcon, W. P., & Naylor, R. L. (2008). Prioritizing climate change adaptation needs for food security in 2030. *Science*, *319*(5863), 607-610.

Masters, R. (2019). *World Meteorological Organization's 2018 Observer Activities report*.

Mi, Z., Guan, D., Liu, Z., Liu, J., Viguié, V., Fromer, N., & Wang, Y. (2019). Cities: The core of climate change mitigation. *Journal of cleaner production*, *207*, 582-589.

Nardone, Alessandro, Bruno Ronchi, Nicola Lacetera, Maria Stella Ranieri, and Umberto Bernabucci. "Effects of climate changes on animal production and sustainability of livestock systems." *Livestock Science* 130, no. 1-3 (2010): 57-69.

Ormazabal, M., Prieto-Sandoval, V., Santos, J., & Jaca, C. (2020). Guiding SMEs towards the Circular Economy: A Case Study. In *Industrial Symbiosis for the Circular Economy* (pp. 27-41).Springer, Cham.

Peterson, L., & Skovgaard, J. (2019). Bureaucratic politics and the allocation of climate finance. *World Development*, *117*, 72-97.

Qureshi, M. I., Qayyum, S., Nassani, A. A., Aldakhil, A. M., Abro, M. M. Q., &Zaman, K. (2019). Management of various socio-economic factors under the United Nations sustainable development agenda. *Resources Policy*, *64*, 101515.

Ramanathan, V., & Carmichael, G. (2008). Global and regional climate changes due to black carbon. *Nature Geoscience*, *1*(4), 221-227.

Rind, D., Perlwitz, J., & Lonergan, P. (2005). AO/NAO response to climate change: 1. Respective influences of stratospheric and tropospheric climate changes. *Journal of Geophysical Research: Atmospheres*, *110*(D12).

Rippke, Ulrike, Julian Ramirez-Villegas, Andy Jarvis, Sonja J. Vermeulen, Louis Parker, Flora Mer, Bernd Diekkrüger, Andrew J. Challinor, and Mark Howden. "Timescales of transformational climate change adaptation in sub-Saharan African agriculture." *Nature Climate Change* 6, no. 6 (2016): 605-609.

Rodriguez, R. S., Ürge-Vorsatz, D., & Barau, A. S. (2018). Sustainable Development Goals and climate change adaptation in cities. *Nature Climate Change*, *8*(3), 181-183.

Salvia, A. L., Leal Filho, W., Brandli, L. L., & Griebeler, J. S. (2019). Assessing research trends related to Sustainable Development Goals: Local and global issues. *Journal of cleaner production*, *208*, 841-849.

Shan, Y., Guan, D., Hubacek, K., Zheng, B., Davis, S. J., Jia, L., & Meng, J. (2018). City-level climate change mitigation in China. *Science advances*, *4*(6), eaaq0390.

Stouffer, Ronald J., Jianjun Yin, J. M. Gregory, K. W. Dixon, M. J. Spelman, W. Hurlin, A. J. Weaver et al. "Investigating the causes of the response of the thermohaline circulation to past and future climate changes." *Journal of Climate* 19, no. 8 (2006): 1365-1387.

Tao, F., Yokozawa, M., Xu, Y., Hayashi, Y., & Zhang, Z. (2006).Climate changes and trends in phenology and yields of field crops in China, 1981–2000. *Agricultural and Forest Meteorology*, *138*(1-4), 82-92.

Trenberth, K. E., Cheng, L., Jacobs, P., Zhang, Y., & Fasullo, J. (2018). Hurricane Harvey links to ocean heat content and climate change adaptation. *Earth's Future*, *6*(5), 730-744.

Wang, G., Bai, W., Li, N., & Hu, H. (2011). Climate changes and its impact on the tundra ecosystem in Qinghai-Tibet Plateau, China. *Climatic Change*, *106*(3), 463-482.

Winkler, H., & Dubash, N. K. (2016). Who determines the transformational change in development and climate finance? *Climate Policy*, *16*(6), 783-791.

Wu, S. H., Yin, Y. H., Zheng, D., & Yang, Q. Y. (2005). Climate changes in the Tibetan Plateau during the last three decades. *Acta Geographica Sinica*, *60*(1), 3-11.

Xu, Q., Xiao, J., Li, Y., Tian, F., & Nakagawa, T. (2010). Pollen-based quantitative reconstruction of Holocene climate changes in the Daihai Lake area, Inner Mongolia, China. *Journal of Climate*, *23*(11), 2856-2868.

Yang, K., Wu, H., Qin, J., Lin, C., Tang, W., & Chen, Y. (2014). Recent climate changes over the Tibetan Plateau and their impacts on energy and water cycle: A review. *Global and Planetary Change*, *112*, 79-91.

Yunev, O. A., Carstensen, J., Moncheva, S., Khaliulin, A., Ærtebjerg, G., & Nixon, S. (2007). Nutrient and phytoplankton trends on the western Black Sea shelf in response to cultural eutrophication and climate changes. *Estuarine, Coastal and Shelf Science*, *74*(1-2), 63-76.

Zazo, C., Dabrio, C. J., Goy, J. L., Lario, J., Cabero, A., Silva, P. G., & Roquero, E. (2008). The coastal archives of the last 15 ka in the Atlantic–Mediterranean Spanish linkage area: Sea level and climate changes. *Quaternary International*, *181*(1), 72-87.

Zhang, Qiang, Marco Gemmer, and Jiaqi Chen. "Climate changes and flood/drought risk in the Yangtze Delta, China, during the past millennium." *Quaternary International* 176 (2008): 62-69.

Zidanšek, Aleksander, Robert Blinc, Anton Jeglič, Skender Kabashi, Sadik Bekteshi, and Ivo Šlaus. "Climate changes, biofuels, and the sustainable future." *International Journal of Hydrogen Energy* 34, no. 16 (2009): 6980-6983.

ABOUT THE EDITORS

Dr. Amit Kumar Thakur

Dr. Amit Kumar Thakur is currently associated with Lovely Professional University as Associate Professor with more than seventeen years of experience in academics and research. His area of expertise includes Renewable energy technologies, Thermal energy, Robotics and Internet of Things. He has organized and conducted a number of workshops, summer internships and expert lectures for students as well as faculty. He has published around Twenty four research papers in referred journals/conferences. He is the Life member of The Indian Society for Technical Education (ISTE), faculty member of Society of Automotive

Engineers (SAE), life member of The Aeronautical Society of India (AMeASI). He is the editorial board member and reviewer of various reputed journals. He was the Research sub-coordinator for ADRDE, DRDO in the project of design and development of the deceleration system for Space capsule recovery experiment (SRE). He has five patents to his name. He is the editor of two books, one from CRC press by Taylor & Francis and one from Nova Science Publishers.

Dr. Rajesh Singh

Dr. Rajesh Singh is currently associated with Lovely Professional University as Professor with more than Sixteen years of experience in academics. He has been awarded as gold medalist in M.Tech from RGPV, Bhopal (M.P) India and honors in his B.E from Dr. B.R. Ambedkar University, Agra (U.P), India. His area of expertise includes embedded systems, robotics, wireless sensor networks and Internet of Things. He has been honored as keynote speakers and session chair to international/ national conferences, faculty development programs and workshops. He has hundred and fifty two patents in his account. He has published more than hundred research papers in referred journals/conferences and twenty four books in the area of Embedded Systems and Internet of Things with reputed publishers like CRC/Taylor & Francis, Narosa, GBS, IRP, NIPA, River Publishers, Bentham Science and RI publication. He is editor to a special issue published by AISC book series, Springer in 2017 & 2018 and IGI global in 2019. Under his mentorship students have participated in national/international competitions including "Innovative Design Challenge competition" by Texas and DST and Laureate award of excellence in robotics engineering, Madrid, Spain in 2014 & 2015. His team has been the winner of "Smart India Hackathon-2019" hardware version conducted by MHRD, Government of India for the problem statement of Mahindra & Mahindra. Under his mentorship students team got "InSc award 2019" under students projects program. He has been awarded with "Gandhian Young Technological Innovation (GYTI) Award", as Mentor to "On Board Diagnostic Data Analysis System-OBDAS", Appreciated under "Cutting Edge Innovation" during Festival of Innovation and Entrepreneurship at Rashtrapati Bahawan, India in 2018. He has been honored with "Certificate of Excellence" from 3rd faculty branding awards-15, Organized by EET CRS research wing for excellence in professional education and Industry, for the category "Award for Excellence in Research", 2015 and young investigator award at the International Conference on Science and Information in 2012.

Dr. Ajay Kumar Kaviti

Dr. Ajay Kumar Kaviti is Associate Professor in Department of Mechanical Engineering, VNR Vignana Jyothi Institute of Engineering & Technology, Hyderabad, India. He did his M.Tech in 2003 and PhD in

2010 from Maulana Azad National Institute of Technology, Bhopal. He has more than sixteen years of experience in teaching and research. He has multi-disciplinary research interests in the areas of renewable energy, alternative fuels and composites. He guided 15 M.Tech projects and currently guiding PhD students. He has 40 publications to his credit in highly reputed international journals along with 20 publications in proceedings of national and international conferences. He is on review board of many prestigious national and international journals and editorial board member for various international journals.

Dr. Anita Gehlot

Dr. Anita Gehlot is currently associated with Lovely Professional University as Associate Professor with more than twelve years of experience in academics. Her area of expertise includes embedded systems, wireless sensor networks and Internet of Things. She has been honoured as keynote speakers and session chair to international/national conferences, faculty development programs and workshops. She has hundred and thirty two patents in her account. She has published more than seventy research papers in referred journals/conferences and *twenty four* books in the area of Embedded Systems and Internet of Things with reputed publishers like CRC/Taylor & Francis, Narosa, GBS, IRP, NIPA,

River Publishers, Bentham Science and RI publication. She is editor to a special issue published by AISC book series, Springer in 2018 and IGI global in 2019. She has been awarded with "*certificate of appreciation*" from University of Petroleum and Energy Studies for exemplary work. Under her mentorship students team got "InSc award 2019" under students projects program. She has been awarded with "Gandhian Young Technological Innovation (GYTI) Award", as Mentor to "On Board Diagnostic Data Analysis System-OBDAS", Appreciated under "Cutting Edge Innovation" during Festival of Innovation and Entrepreneurship at Rashtrapati Bahawan, India in 2018.

Dr. JV Muruga Lal Jeyan

Dr. JV Muruga Lal Jeyan is associated with Aerospace engineering of LPU. His recent research areas are High speed aerodynamics, flight instrumentation and superconductivity electric aircrafts. He is the former SBDE and elected as CEO of Economic international Singapore for 3 years. Published more than 30 International articles. Holds 13 plus years of experience in the field of academic, industry and research. Has served 9 plus years as Head of Aviation Studies in various Universities. Has an excellent number of research scholar's and students working under him in various research areas in and around India. Received *Rashtriya Gaurav*

Award for meritorious service, by former governor of Tamilnadu & Assam. Life member of 1) Aeronautical society of India 2) IAETSD 3) The institute of engineers IE 4) The Indian Science congress ISC 5) SAISE. He is also serving as Editorial Board Member for few international reputed journals. He has organized and chaired more than 15 international and national events. He has visited 8 plus country and 100 plus universities around the globe. Recently he has been elected as organizing chair for aerospace stream of 2nd world summit Indiana 2019 to be held at USA.

INDEX

A

B

C

D

E

F

G

H

I

L

Q

R

S

T

U

V

W